I0503964

United States Department of Agriculture

Economic Research Service
www.ers.usda.gov

Access this report online:

www.ers.usda.gov/publications/eib-economic-information-bulletin/eib124.aspx

Download the charts contained in this report:

- Go to the report's index page www.ers.usda.gov/publications/ eib-economic-information-bulletin/eib124.aspx
- Click on the bulleted item "Download eib124.zip"
- Open the chart you want, then save it to your computer

Recommended citation format for this publication:

Fernandez-Cornejo, Jorge, Richard Nehring, Craig Osteen, Seth Wechsler, Andrew Martin, and Alex Vialou. *Pesticide Use in U.S. Agriculture: 21 Selected Crops, 1960-2008*, EIB-124, U.S. Department of Agriculture, Economic Research Service, May 2014.

Cover images: Shutterstock.

Use of commercial and trade names does not imply approval or constitute endorsement by USDA.

United States Department of Agriculture

Economic Research Service

Economic Information Bulletin Number 124

May 2014

Pesticide Use in U.S. Agriculture: 21 Selected Crops, 1960-2008

Jorge Fernandez-Cornejo, Richard Nehring, Craig Osteen, Seth Wechsler, Andrew Martin, and Alex Vialou

Abstract

Pesticide use has changed considerably over the past five decades. Rapid growth characterized the first 20 years, ending in 1981. The total quantity of pesticides applied to the 21 crops analyzed grew from 196 million pounds of pesticide active ingredients in 1960 to 632 million pounds in 1981. Improvements in the types and modes of action of active ingredients applied along with small annual fluctuations resulted in a slight downward trend in pesticide use to 516 million pounds in 2008. These changes were driven by economic factors that determined crop and input prices and were influenced by pest pressures, environmental and weather conditions, crop acreages, agricultural practices (including adoption of genetically engineered crops), access to land-grant extension personnel and crop consultants, the cost-effectiveness of pesticides and other practices in protecting crop yields and quality, technological innovations in pest management systems/practices, and environmental and health regulations. Emerging pest management policy issues include the development of glyphosate-resistant weed populations associated with the large increase in glyphosate use since the late 1990s, the development of Bt-resistant western corn rootworm in some areas, and the arrival of invasive or exotic pest species, such as soybean aphid and soybean rust, which can influence pesticide use patterns and the development of Integrated Pest Management programs.

Acknowledgments

The authors thank Marc Ribaudo and James MacDonald, USDA, Economic Research Service; Sheryl Kunickis, David Epstein, Julius Fajardo, and Teung Chin, USDA, Office of Pest Management Policy; Cynthia Doucoure, U.S. EPA; Scott Swinton, Michigan State University; and David Shaw, Mississippi State University. We are grateful to Dale Simms for valuable editorial assistance and Cynthia A. Ray for graphics and layout.

Contents

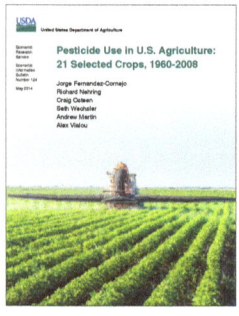

USDA

United States Department of Agriculture

A report summary from the Economic Research Service   May 2014

Pesticide Use in U.S. Agriculture: 21 Selected Crops, 1960-2008

Jorge Fernandez-Cornejo, Richard Nehring, Craig Osteen, Seth Wechsler, Andrew Martin, and Alex Vialou

Find the full report
at *www.ers.usda.
gov/publications/eib-
economic-information-
bulletin/eib-124.aspx*

What Is the Issue?

Pesticides—including herbicides, insecticides, and fungicides—have contributed to substantial increases in crop yields over the past five decades. Properly applied, pesticides contribute to higher yields and improved product quality by controlling weeds, insects, nematodes, and plant pathogens. In addition, herbicides reduce the amount of labor, machinery, and fuel used for mechanical weed control. However, because pesticides may possess toxic properties, their use often prompts concern about human health and environmental consequences. The examination of pesticide use trends is critical for informed pesticide policy debate and science-based decisions. This report analyzes pesticide use trends using a new pesticide database compiled from USDA and proprietary data, focusing on 21 crops.

What Did the Study Find?

Total pesticide use, as well as the specific active ingredients used (for example, with novel target sites of action or improved toxicological profiles), has changed considerably over the past five decades.

Pesticide use on the 21 crops analyzed in this report rose rapidly from 196 million pounds of active ingredient (a.i.) in 1960 to 632 million pounds in 1981, largely because of the increased share of planted acres treated with herbicides to control weeds. In addition, the total planted acreage of corn, wheat, and, in particular, soybeans increased from the early 1960s to early 1980s, which further increased herbicide use. Most acres planted with major crops (particularly corn and soybeans) were already being treated with herbicides by 1980, so total pesticide use has since trended slightly downward driven by other factors, to 516 million pounds in 2008 (the most recent year for which we have enough complete data).

The rapid adoption of herbicides was mainly driven by relative price declines that helped reduce the cost of herbicides relative to other pest control practices and encouraged substitution of herbicides for labor, fuel, and machinery use in mechanical weed control. The fluctuations in pesticide use over 1982-2008 were driven by several factors, including changes in planted acreage, crop and input prices, weather, pesticide regulations, and the introduction of new pesticides and genetically engineered (GE) seed. Changes in the acreages of corn, cotton, soybeans, potatoes, and wheat contributed to fluctuations in pesticide use from 1981 to 2008, with many high and low years in herbicide and pesticide use coinciding with high and low years in total acreage of these crops.

The pesticide types applied by U.S. farmers for the 21 crops analyzed changed considerably from 1960 to 2008. ***Insecticides*** accounted for 58 percent of pounds applied in 1960, but only 6 percent in 2008. On the other hand, ***herbicides*** accounted for 18 percent of the pounds applied in 1960 but 76 percent by 2008. The growth of herbicide use is also illustrated by the percent of acres treated.

ERS is a primary source of economic research and analysis from the U.S. Department of Agriculture, providing timely information on economic and policy issues related to agriculture, food, the environment, and rural America.

www.ers.usda.gov

Approximately 5-10 percent of corn, wheat, and cotton acres were treated with herbicides in 1952. By 1980, herbicide use had reached 90-99 percent of U.S. corn, cotton, and soybean acres planted. Notably, the four most heavily used active ingredients in 2008 (glyphosate, atrazine, acetochlor, and metolachlor) were all herbicides. *Fungicides'* share of pesticide use has remained at 7 percent or less since 1971, down from 11-13 percent in the early 1960s. *Other pesticides*—which include soil fumigants, desiccants, harvest aids, and plant growth regulators—generally accounted for 5-11 percent of total pesticide use from 1960 to 1992, increased to 17 percent of use in 2002, and then declined to 13 percent in 2008.

Total **pesticide expenditures** in U.S. agriculture reached close to $12 billion in 2008, a 5-fold increase in real terms (adjusted for inflation) since 1960, but well below the $15.4-billion peak reached in 1998.

In 2008, corn, soybeans, cotton, wheat, and potatoes accounted for about 80 percent of the pesticide quantity (measured in pounds of a.i.) applied to the 21 crops examined. *Corn* has been the top pesticide-using crop in the United States since 1972 and received about 39 percent of the pesticides in 2008 (mostly herbicides). While corn is a major component of livestock feed, expansion of ethanol production for fuel use has boosted corn acres in recent years. The increase in corn acreage led to an increase in pesticide use and change in the active ingredients used. The change in active ingredients also reflects increased glyphosate use associated with the adoption of HT crops.

Soybean production had the next largest share in 2008 (22 percent), almost all of which were herbicides. *Potatoes'* share rose significantly in the 1990s and reached about 10 percent by 2008. Other pesticides, including soil fumigants and desiccants, constituted a large portion of the pesticides applied to potatoes in 2008. *Cotton* accounted for just over 7 percent of the pesticides, mostly insecticides, in 2008, a major reduction from its 40-percent share in the early 1960s. The quantity applied to cotton trended downward since 1972 due to the replacement of DDT and other older insecticides with more effective products, eradication of the boll weevil, and adoption of Bt cotton. *Wheat* accounted for less than 5 percent of the pesticides, mostly herbicides, in 2008.

How Was the Study Conducted?

The study analyzes a new pesticide database that was compiled from pesticide use surveys carried out by USDA's National Agricultural Statistics Service (NASS) and Economic Research Service (ERS), supplemented by proprietary data provided by a market research company to the U.S. Environmental Protection Agency (EPA), and shared with ERS under an agreement between the two agencies.

The data were collected for 1960-2008 and focus on 21 crops: apples, barley, corn, cotton, grapefruit, grapes, lemons, lettuce, peaches, peanuts, pears, pecans, potatoes, oranges, rice, sorghum, soybeans, sugarcane, sweet corn, tomatoes, and wheat. These crops account for roughly 72 percent of total conventional pesticide use in U.S. agriculture. This report discusses "conventional" pesticides defined by the EPA as substances developed and produced primarily or only for use as pesticides and excludes sulfur, petroleum distillate, sulfuric acid, and hydrated lime. In addition to data described above, the study used pesticide expenditure data covering all U.S. agriculture drawn from ERS publications.

Pesticide use by crop, 21 selected crops, 2008, percent total pounds of active ingredient applied

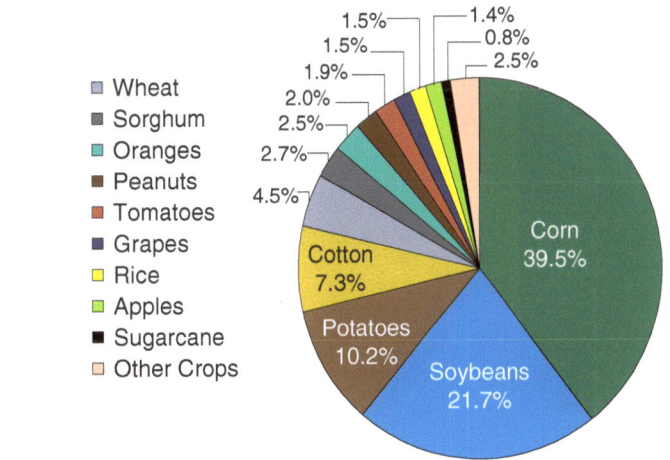

Note: "Other Crops" include: lettuce, pears, sweet corn, barley, peaches, grapefruit, pecans, and lemons.
Sources: Economic Research Service with USDA and proprietary data. See Appendix 2.

www.ers.usda.gov

Pesticide Use in U.S. Agriculture: 21 Selected Crops, 1960-2008

Introduction

Prior to World War II, farmers managed pests using cultural practices and a few inorganic pesticides (see box, "Agricultural Pests and Pesticides"). After World War II, new, synthetic organic materials—such as the insecticide DDT and the herbicide 2,4-D—enhanced farmers' pest control options. These pesticides made crop production more efficient by providing superior crop protection and reducing the need for tillage (Padgitt et al., 2000).

Pesticides, together with fertilizers and improved seed varieties, have contributed to substantial increases in crop yields over the last 80 years. Average corn yields rose from 20 bushels per acre in 1930 to more than 150 bushels per acre in recent years. During the same period, cotton yields rose nearly fourfold, and soybean yields increased more than threefold (Fernandez-Cornejo, 2004).

Pesticides are used to prevent or manage pests such as weeds, insects, and plant pathogens, while reducing the amount of labor, fuel, and machinery used for pest control (Osteen and Szmedra, 1989; Fernandez-Cornejo et al., 1998; Gardner et al., 2009). These benefits translate into lower production costs, higher crop yields and/or quality, and increased profits for farmers. The benefits for U.S. farmers are evidenced by their willingness to spend approximately $12 billion on pesticides in 2008 (USDA/ERS, 2010b). Many consumers also benefit from abundant, and relatively inexpensive, unblemished foods (Fernandez-Cornejo et al., 1998).[1] However, not all consumers feel that the benefits of heavier reliance on pesticide use outweigh the costs, accounting for the recent growth in organically grown food sales.

The benefits of pesticide use are accompanied by potential risks to human health and the environment. Human health risks can result from direct exposure of farm workers to pesticides or from consumer exposure to pesticide residues on foods.[2] Environmental risks can result from the movement of pesticides into ground and surface water and into the food chain (Council of Environmental Quality, 1993).

By the 1960s, concerns about wildlife and human health led to calls for more stringent pesticide regulation. In 1972, Congress empowered the U.S. Environmental Protection Agency (EPA) to review the safety of existing pesticides (table 1). The EPA determined that some pesticides, such as DDT, aldrin, dieldrin, chlordane, and heptachlor, posed unreasonable risks and their registrations were subsequently canceled. Other compounds faced rigorous scrutiny in the 1990s and 2000s, as the EPA required additional studies of individual chemicals' toxicity and focused on the human health risks associated with pesticide residues (see Appendix 1, "Human Health Effects and Pesticide Regulation").

---

[1] By controlling insects, diseases, and other pests, pesticides can provide less costly food that is also free from harmful organisms and blemishes.

[2] The EPA sets pesticide residue limits (known as tolerances) on food to protect consumers from harmful levels of pesticides (EPA, 2008).

Agricultural Pests and Pesticides

From the point of view of agriculture, **pests** are "organisms that diminish the value of resources in which man is interested as they interfere with the production and utilization of crops and livestock" used for food and fiber (NRC, 1975). The term pest includes insects, mites, nematodes, plant pathogens, weeds, and vertebrates. Pests can reduce crop yields or quality of production, while costs of managing pests increase production costs.

The term **pesticide** includes the substances used to control pests. It includes herbicides (to control weeds and other plants), insecticides (to control insects), fungicides (to control fungi or other plant pathogens), nematicides (to control parasitic worms), and rodenticides (to control rodents). The term pesticide also encompasses soil fumigants, plant growth regulators, defoliants, and desiccants. Pesticides can be synthetic (developed in laboratories and manufactured) or natural. According to a study conducted using 1996 data (Fernandez-Cornejo and Jans, 1999), weeds are by far the most important pests in U.S. agriculture in terms of the share of treatments used to control them.

The active chemicals used to control pests (the biologically active part of the pesticide) are called pesticide **active ingredients**. Pesticides are sold as mixtures of these active ingredients with inert materials used to improve safety and facilitate storage, handling, or application. Appendix table 1.2 provides a list of the pesticide active ingredients included in this report. All pounds of pesticides referred to in this report are in terms of active ingredients.

The term pesticide used in this report includes what EPA defines as "conventional pesticides" (EPA, 2011). This means pesticides that are chemicals or other substances developed and produced primarily or only for use as pesticides. It excludes sulfur, petroleum oil and other chemicals used as pesticides (for example, sulfuric acid, hydrated lime, and insect repellents).

This report examines trends in pesticide use in U.S. agriculture from 1960 to 2008 focusing on 21 crops that account for more than 70 percent of pesticide use and identifies the factors affecting these trends. The report also provides a detailed analysis of trends for the five crops that represent the largest users of pesticides: corn, cotton, potatoes, soybeans, and wheat.

Data

The data were compiled from pesticide use surveys carried out by USDA's Economic Research Service (ERS) and National Agricultural Statistics Service (NASS), as well as from proprietary data provided by a market research company to the Environmental Protection Agency (EPA), and shared with ERS under an agreement between the two agencies (proprietary data for short). This report only uses published EPA estimates of aggregate pesticide use for comparison, because the EPA does not publish estimates of pesticide use for individual crops, as this report does.

The ERS surveys covered various crops and were conducted mainly in the 1960s and 1970s (Eichers et al., 1968, Eichers et al., 1970, Andrilenas, 1974; Eichers et al., 1978, Delvo et al., 1983; USDA/ERS, 1984). The NASS pesticide surveys (USDA/NASS, various years) began in 1990, but not all crops were covered every year.

Table 1
Basic Pesticide Legislation

The Insecticide Act of 1910 – Prohibited the manufacture, sale, or transport of adulterated or misbranded pesticides; protected farmers and ranchers from marketing of ineffective products.

Federal Food, Drug, and Cosmetic Act of 1938 (FFDCA) – Provided that safe tolerances be set for residues of unavoidable poisonous substances, such as pesticides, in food.

Federal Insecticide, Fungicide, and Rodenticide Act of 1947 (FIFRA) – Required pesticides to be registered before sale and the product labeled to specify content and whether the substance was poisonous.

Miller Amendment to FFDCA of 1954 – Amended the Federal Food, Drug, and Cosmetic Act (FFDCA) to require that tolerances for pesticide residues be established (or exempted) for food and feed (Section 408). Allowed consideration of risks and benefits in setting tolerances.

Food Additives Amendment to FFDCA of 1958 – Amended FFDCA to give authority to regulate food additives against a general safety standard that does not consider benefits (Section 409); included the Delaney Clause prohibiting food additives found to induce cancer in humans or animals. Pesticide residues in processed foods were classified as food additives, while residues on raw commodities were not. When residues of a pesticide applied to a raw agricultural commodity appeared in a processed product, the residues in processed foods were not to be regulated as food additives if levels were no higher than sanctioned on the raw commodity.

FIFRA Amendments of 1964 – Increased authority to remove pesticide products from the market for safety reasons by authorizing denial or cancellation of registration and the immediate suspension of a registration, if necessary, to prevent an imminent hazard to the public.

Federal Environmental Pest Control Act (FEPCA) of 1972 – Amended FIFRA to significantly increase authority to regulate pesticides. Allowed registration of a pesticide only if it did not cause "unreasonable adverse effects" to human health or the environment; required an examination of the safety of all previously registered pesticide products within 4 years using new health and environmental protection criteria. Materials with risks that exceeded those criteria were subject to cancellation of registration. Specifically included consideration of risks and benefits in these decisions.

FIFRA Amendment of 1975 – Required consideration of the effects of registration cancellation or suspension on the production and prices of relevant agricultural commodities.

Federal Pesticide Act of 1978 – Identified review of previously registered pesticides as reregistration; eliminated the deadline for reregistration but required an expeditious process.

FIFRA Amendments of 1988 – Accelerated the reregistration process by requiring that all pesticides containing active ingredients registered before November 1, 1984, be reregistered by 1995; provided EPA with additional financial resources through reregistration and annual maintenance fees levied on pesticide registrations.

Food Quality Protection Act of 1996 (FQPA) – Amended FIFRA and FFDCA to set a consistent safety standard for risks from pesticide residues in foods: "ensure that there is a reasonable certainty that no harm will result to infants and children from aggregate exposure." Pesticide residues are no longer subject to the Delaney Clause of FDCA; both fresh and processed foods may contain residues of pesticides classified as carcinogens at tolerance levels determined to be safe. EPA was required to reassess existing tolerances of pesticides within 10 years, with priority to pesticides that may pose the greatest risk to public health. Benefits no longer have a role in setting new tolerances, but may have a limited role in decisions concerning existing tolerances. Included special provisions to encourage registration of minor-use and public health pesticides.

Pesticide Registration Improvement Act of 2003 – Amended FIFRA to provide for the enhanced review of covered pesticide products, to authorize service fees for registration actions in the Antimicrobials, Biopesticides and Pollution Prevention, and Registration Divisions of EPA's Office of Pesticide Programs.

The dataset aggregates pesticides applied by year, State, crop, and active ingredient (a.i.). When both USDA and proprietary data are available, ERS and NASS data are used. Proprietary data are used when USDA data are not available. If neither USDA data nor proprietary data are available for a specific year, crop, State, and active ingredient, estimates are made based on application rates (e.g., pounds of a.i. per acre) from contiguous years and planted acres reported by USDA (see appendix table 2.1 for a list of main sources). ERS did not have access to the proprietary data needed to estimate pesticide quantities beyond 2008 using this method.

The 21 (7 major and 14 minor) crops included in this report —apples, barley, corn, cotton, grapefruit, grapes, lemons, lettuce, peaches, peanuts, pears, pecans, fall potatoes, oranges, rice, sorghum, soybeans, sugarcane, sweet corn, tomatoes, and wheat—account, on average, for 72 percent of the total conventional pesticide use in U.S. agriculture (including all crops) as estimated by EPA from 1964 through 2007 (EPA, 1999, 2011). Additional information is included in Appendix 2.

This report also contains charts showing trends in the share of acreages treated with major pesticide types for cotton, corn, soybeans, wheat, and potatoes to provide insight into factors influencing trends in pesticide quantities. These charts only include published NASS estimates and linear interpolations between published estimates for years when NASS estimates were not available, because the estimates were difficult to obtain from the proprietary data.

In addition to data described above, the study used pesticide expenditure data covering all U.S. agriculture developed for annual farm income accounts (USDA/ERS, 2010b). Additionally, a set of physical characteristics was obtained for nearly 200 of the active ingredients (appendix table 2.2) used in corn, cotton, sorghum, and soybean production (Wauchope et al., 1992; Kellogg et al., 2002) to illustrate the estimation of quality-adjusted indices for pesticide prices and quantities for four major crops.[3]

This report discusses conventional pesticides, defined by EPA as "chemicals or other substances developed and produced primarily or only for use as pesticides" (EPA, 2011). Conventional pesticides exclude chemicals "that are produced and marketed mostly for other purposes (i.e. multi-use chemicals)." Notably, this report excludes sulfur, petroleum distillate products, sulfuric acid, and hydrated lime. This report maintains consistency with previous ERS reports and comparability of these 1960-2008 estimates with the EPA 1964-2007 agricultural pesticide use estimates.[4] (Unlike this report, which focuses on crop-specific pesticide use, EPA reports estimates of pesticide use for the whole agricultural sector.) Previous ERS pesticide use reports either separated sulfur and petroleum use estimates from conventional pesticides, or did not report their use; ERS summaries of national pesticide use surveys for 1964, 1966, 1971, and 1976 separated sulfur and petroleum use from conventional pesticide use, while the 1982 summary did not report their use.

Among previous USDA pesticide use reports, Osteen and Szmedra (1989) discussed trends in agricultural chemical use through 1982; Lin et al. (1995) discussed trends for 11 crops from 1964 through 1992, and Livingston and Osteen (2012) provided a brief summary of pesticide use for 5 major crops using NASS data, all excluding sulfur and petroleum in their pesticide use summaries.

---

[3]Pesticide quality has changed as materials more effective and less harmful to human health and the environment have been introduced while others have been banned or dropped by their manufacturers (Fernandez-Cornejo and Jans, 1995).

[4]EPA estimates also exclude pesticides on treated seed or applied as seed treatments, because neither the USDA nor the proprietary data include this information. While large acreages of corn, cotton, and soybeans may be planted with treated seed, the contribution to aggregate pesticide quantity is small.

Similarly, EPA's 1964-2007 series of national estimates of conventional pesticide use excludes sulfur and petroleum, as well as other chemicals used as pesticides (such as sulfuric acid and insect repellants), wood preservatives, specialty biocides, and chlorine/hypochlorites.[5]

---

[5]EPA publishes a time series that includes sulfur, petroleum, and other non-conventional pesticides, but only for 1979-2007 and no breakdown by crop.

Economic Factors Influencing Pesticide Use

Pesticide patterns, including changes in aggregate pesticide use and active ingredients applied, have been influenced by a number of factors over the last 50 years. Major factors affecting the demand and/or the supply of pesticides include:

- *Pest infestation levels*. Commonly, at low densities, a pest causes little or no damage to agriculture, but potentially pests "can erupt to population densities that cause devastations of entire crops" (NRC, 1989, p. 27). From an economic viewpoint, an agricultural pest is an "animal or plant pathogen whose population density exceeds some unacceptable threshold level, resulting in economic damage" (Horn, 1988).

- *Technical improvements* that increase pesticide effectiveness and/or reduce environment/ human health consequences. For example, the development of new chemistries for herbicides such as the very active sulfonylureas and imidazolinones in the 1980s and 1990s reduced the application rates from multiple pounds per acre to a few ounces, or even fractions of an ounce. Similarly, low-use-rate insecticide compounds were introduced, such as synthetic pyrethroids (permethrin, cypermethrin) in the mid-1970s and neo-nicotinoids (imidicloprid, clothianidin) in the mid-1990s, increasing effectiveness.

- The use of *Integrated Pest Management (IPM)* practices by U.S. farmers, including crop rotations, mixing or alternating pesticides to reduce development of pest resistance to pesticides, biological monitoring of pests to better understand population densities and phenological development, predictive models that use local weather conditions along with pest phenology to optimize spray timings, improved and more accurate spray technologies, the optimization of planting and harvest dates, and the use of beneficial insects (Fernandez-Cornejo and Jans, 1999).

- The adoption of *genetically engineered (GE) crops*, including herbicide-tolerant corn, soybeans, and cotton, as well as insect-resistant corn and cotton (Fernandez-Cornejo and Caswell, 2004). GE seeds have genes that provide specific traits such as herbicide tolerance (HT) and insect resistance. HT crops tolerate potent herbicides, allowing adopters of these varieties to control pervasive weeds more effectively. Insect-resistant (Bt) crops contain genes from the soil bacterium *Bacillus thuringiensis* that produce a protein toxic to specific insects, protecting the plant over its entire life (Fernandez-Cornejo and McBride, 2002).

- The use of *conservation practices* that yield environmental benefits and reduce soil erosion but may or may not increase pesticide use (Fuglie, 1999). For example, the adoption of conservation tillage by U.S. soybean growers rose from about 30 percent in 1996 to 63 percent in 2006 (Fernandez-Cornejo et al., 2012).[6] Glyphosate use has been associated with use of conservation tillage as it replaced more intensive tillage.

- *Pesticide regulations*. For example, the ban of DDT forced growers to use different insecticides and pesticide producers to develop new crop protection products (Osteen, 1987).

---

[6]Conservation tillage includes no-till and reduced till. No-till is often considered the most effective of all conservation tillage systems. Soybean farmers adopted no-till in 45 percent of their planted acres in 2006 (Horowitz et al., 2010).

- *Changes in crop acreage*. Aggregate pesticide use has increased during periods of increasing planted acreage, and decreased during periods of decreasing acreage. Acreage planted reflects farm policy and economic factors that influence crop demand. In particular, changes in acreage under diversion programs and the Conservation Reserve Program (CRP) influenced the acreage in crop production and thus pesticide use.[7]

- *Increased corn acreage*. While corn is a major component of livestock feed, expansion of ethanol production for fuel use since the early 2000s, reflecting energy policy, boosted corn acres even further (Westcott, 2007). The increase in corn acreage led to an increase in pesticide use and change in the active ingredients used. The change in active ingredients also reflects increased glyphosate use associated with the adoption of HT crops.

In addition, many factors indirectly influence pesticide use, such as weather, cultivar susceptibility or resistance to pests or disease, and consumer demand for blemish-free produce that encourages growers to apply pesticide materials even when their IPM monitoring indicates otherwise. Other factors include commodity prices as well as the prices of other inputs such as labor and machinery.

Farmers' Choice of Pesticides

According to economic efficiency criteria, farmers choose the combination of pest control practices that maximizes the difference between pest damage reductions and control costs (Osteen and Szmedra, 1989).

In contrast to other inputs such as fertilizers, capital, and labor, which affect yields directly, pesticides have an indirect effect on yields by reducing crop losses (NRC, 2002). The damage control framework developed by Lichtenberg and Zilberman (1986) recognizes this. This framework assumes that the observed yield (effective yield) is equal to the potential yield (that would obtain in the absence of the damage caused by pests) minus the losses caused by damaging agents (pests). Yield losses from pests are affected by the pest infestation levels (sometimes called pest density) and by the effectiveness of the pest control inputs in managing the infestation (see Appendix 3, "The Economics of Pesticide Use").

Relative to yields prevailing at the time, estimates obtained in the 1970s and 1980s found that the expected losses from insects and plant pathogens without the use of insecticides and fungicides ranged between 1 percent and 26 percent for large-acreage crops like corn, soybeans, and wheat. Peanuts, fruits, and vegetables were estimated to have higher yield losses (Fernandez-Cornejo et al., 1998). Crop losses from weeds not treated with herbicides ranged from 0 to 53 percent for the crops studied. Losses from all pests without pesticides ranged from 24 to 79 percent.

Older estimates reported by the U.N. Food and Agriculture Organization (FAO, 1975) estimated global crop losses to pests at around 35 percent. As Yudelman et al. (1998) reported, FAO "estimated that preharvest losses in developing countries were around 40 percent, while postharvest losses added a further 10 to 20 percent." More recently, Oerke (2006) estimated that, without pest control, production would decline worldwide by 54 percent for corn (maize), 46 percent for soybeans,

---

[7]Historically, economists argued that price and income support and acreage diversion programs encouraged more intensive pesticide use per acre, but changes in farm legislation since 1977 have decreased those incentives (Osteen and Fernandez-Cornejo, 2013). Also, some economists have argued that crop insurance could discourage pesticide use, but empirical evidence is mixed. There is some evidence that crop insurance, including subsidized premiums, encourages producers to grow higher value crops and increase pesticide use (Wu, 1999; Claassen et al., 2011).

75 percent for cotton, 58 percent for potatoes, and 30 percent for wheat, relative to production prevailing in 2001-03.[8] Still, with the exception of controlled experiments, in most cases the estimation of the impact of a treatment is difficult, as they would require us to observe the counterfactual. Thus, caution should be exercised when considering these estimates.

Pesticide Expenditures

Pesticide expenditures are correlated with pesticide use. Expenditures for all pesticides used in U.S. agriculture expressed in nominal terms increased steadily through most of the last half-century. In real terms (adjusted for inflation) expenditures increased from $2.3 billion in 1960 to about $10 billion in the 1980s. Expenditures peaked at $15.4 billion in 1998 before falling to approximately $12 billion in 2008 (see box, "Pesticide Expenditures").

The cost share of pesticides (relative to all input costs) peaked at 4.0 percent in 1998, up from 0.7 percent in 1960 and 1.6 percent in 1970 (see box, "Pesticide Expenditures"). This trend reflects the rapid growth in pesticide use in this period. Between 1998 and 2008, the pesticide cost share declined to about 3.1 percent, reflecting a slowdown in the rate of increase in pesticide prices. Even when pesticide use is high, as in the peak year of 1998, pesticides represent a minor cost component in U.S. agriculture. That year, labor accounted for 20.4 percent, fertilizer for 4.5 percent, land for 12.5 percent, and capital for 12.1 percent of agricultural costs (box table 2).

---

[8]Oerke (2006) also estimated production losses relative to attainable production, which consisted of actual losses that would occur with current pest control practices, additional losses without pest control, and remaining production without pest control. So, actual production was attainable yield minus actual (or current) losses. Oerke estimated that production would decline without pest control practices, relative to attainable production in 2001-03, by 37 percent for corn (maize), 34 for soybeans, 53 percent for cotton, 35 percent for potatoes, and 22 percent for wheat. The estimates did not account for any producer adjustments in response to changing costs and market prices.

Pesticide Expenditures

Nominal expenditures on all pesticides used in U.S. agriculture (box fig. 2.1), increased steadily through most of the last half-century. In real terms (constant 2008 dollars, adjusted for inflation), pesticide expenditures increased five-fold between 1960 and 2008. However, 2008 expenditures remain well below the 1998 peak (in real terms).

Box figure 2.1
Pesticide expenditures in U.S. agriculture, 1960-2008

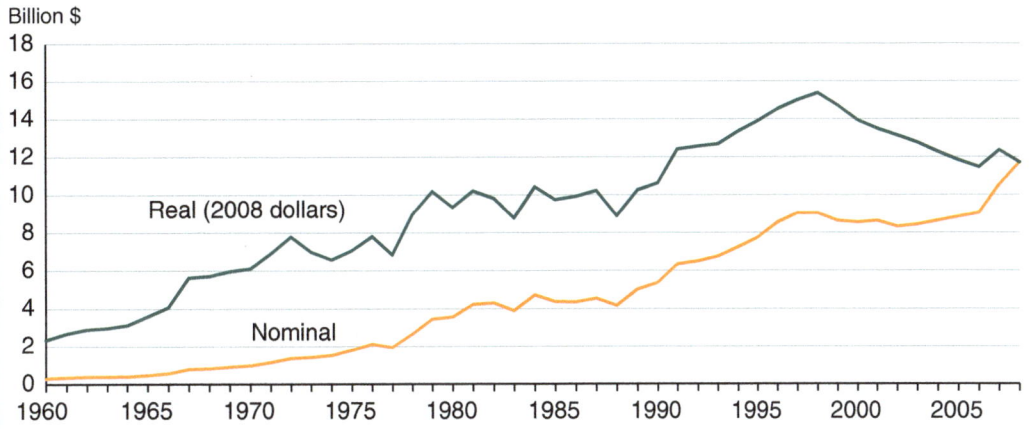

Source: Data from USDA/ERS (2010b). Deflator: Index of Prices Paid by Farmers from USDA/NASS' *Agricultural Prices Summary* (various years).

Per-acre pesticide expenditure s vary widely, generally increasing with the per acre value of the crop. For example, while corn and soybean farmers spend between $17 and $26 per acre, cotton farmers spend more than $65 per acre, and producers of potatoes (a high-value commodity) spend nearly $200 per acre (box table 2.1). Pesticide expenditures for many fruits and vegetables are even higher—$842 per acre for tomatoes and $1,588 per acre for strawberries in 1994 (Fernandez et al., 1998).

Box table 2.1
Value of the average product of pesticides for selected crops and years

Item	Corn 2001	Corn 2005	Potatoes 2008	Soybeans 2002	Cotton 2003	Sorghum 2003	Wheat 2004
Pesticide expenditure, $ per acre	26.44	22.84	193.62	17.12	65.81	17.32	22.84
Yield, unit per acre (bu, lb, cwt)[1]	144	149	395	40	742	47	39.8
Price, $ per unit	1.84	1.74	7.00	5.20	0.66	2.25	3.44
Total revenues, $ per acre	265	259	2,765	208	490	106	137
Average value product, $ of revenue per $ of pesticide expenditure	10.02	11.35	14.28	12.14	7.44	6.11	5.99

[1]Bushels for corn, soybeans, sorghum and wheat; cwt (100 pounds) for potatoes, lb (pounds) for cotton.

Source: ERS analysis of selected USDA Agricultural Resource Management Surveys (ARMS), Cost of Production Surveys and Costs and Returns Reports and proprietary pesticide data.

At the national level, the **cost share of pesticides** (relative to all input costs, box table 2.2) peaked at 4.0 percent in 1998, up from 0.7 percent in 1960 and 1.6 percent in 1970. This trend reflects the rapid growth in pesticide use, particularly in herbicide applications on major field crops, since 1960 (see Appendix 1). Between 1998 and 2008, the pesticide cost share declined to about 3.1 percent, reflecting a slowdown in the rate of increase in pesticide prices. Thus, even when pesticide use is high, as in the peak year of 1998, pesticides represent a minor cost component in U.S. agriculture. That year, labor accounted for 20.4 percent, fertilizer for 4.5 percent, land for 12.5 percent, and capital for 12.1 percent of agricultural costs (box table 2.2). Other intermediate inputs, such as fuel and feed, account for the remaining cost shares.

Box table 2.2

Cost shares in U.S. agriculture (percent)

				Intermediate inputs		
Year	Labor	Capital	Land	All intermediate Inputs[1]	Pesticides	Fertilizer
2008	16.1	8.3	18.8	58.5	3.1	5.8
2005	21.7	9.6	14.7	54.0	3.2	4.2
2000	22.3	13.4	8.8	55.5	3.7	3.8
1998	20.4	12.1	12.5	55.0	4.0	4.5
1990	17.3	14.0	19.0	49.8	2.6	4.1
1985	15.1	20.1	14.8	50.0	2.5	5.7
1970	21.5	12.9	15.2	50.4	1.6	3.8
1960	21.8	10.0	19.2	49.1	0.7	2.9

[1]Intermediate inputs include pesticides, fertilizers, fuels, seeds, and other materials.

Source: ERS calculations from the productivity accounts (USDA/ERS, 2010).

Trends in Pesticide Use

Pesticide use has changed considerably over the past five decades (fig. 1; table 2). Rapid growth in pesticide use characterized the first two decades. The total quantity of pesticides applied to the 21 crops analyzed in this study grew from 196 million pounds of pesticide active ingredients in 1960 to 632 million pounds in 1981. Changes in the active ingredients applied and a slight downward trend in use since 1981 (with some fluctuations in total pesticide use) caused pesticide use totals to dip to 468 million pounds in 1987 before increasing to 601 million pounds in 1997, and ending at 516 million pounds in 2008.

This pattern of pesticide use for the 21 crops parallels the total use of conventional pesticides for U.S. agriculture as estimated by the U.S. Environmental Protection Agency (EPA, 1999, 2011). On average, the amount of pesticide used on the 21 crops included in this study represents 72 percent of the total estimated by EPA for U.S. agriculture (fig. 2).

Early Growth Driven by Herbicide Adoption for Major Crops

Pesticide use more than tripled between 1960 and 1981. Herbicide use increased more than tenfold (from 35 to 478 million pounds) as more U.S. farmers began to treat their fields with these chemicals. By contrast, insecticide use declined from 114 million pounds in 1960 to 97 million pounds in 1981, and fungicide use increased only slightly (from 25 to 27 million pounds).

While farmers have used insecticides and fungicides for many years, the widespread use of herbicides is a more recent phenomenon, as weed control was previously achieved by cultivation and other methods. Only 10 percent of U.S. corn acres planted were treated with herbicides in 1952. By 1976, herbicide use had grown to 90 percent of corn acres planted. Growth slowed in subsequent years, reaching 95 percent in 1982 before stabilizing at around 98 percent in recent years (fig. 3). Herbicides applied to other major crops like soybeans and cotton experienced similar growth patterns (figs. 4-7). For example, approximately 5 percent of cotton acres were treated in 1952 while

Figure 1
Pesticide use in U.S. agriculture, 21 selected crops, 1960-2008

Million pounds of pesticide active ingredient

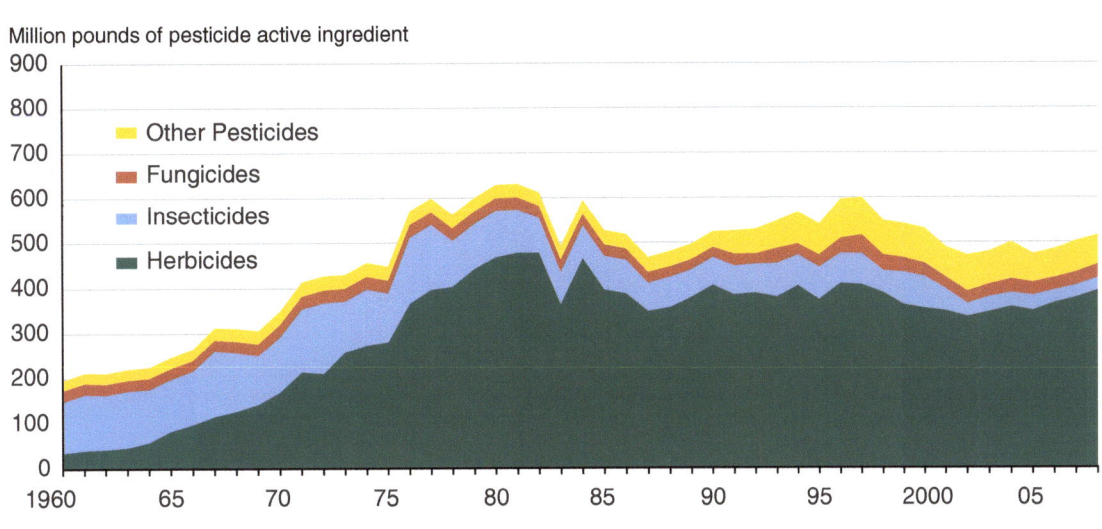

Source: Economic Research Service with USDA and proprietary data. See Appendix 2.

Pesticide Use in U.S. Agriculture: 21 Selected Crops, 1960-2008, EIB-124
Economic Research Service/USDA

Table 2
Pesticide use by crop and type, 21 selected crops, 1960-2008

Crop	1960	1965	1970	1975	1980	1985	1990	1995	2000	2005	2008
					Millions of pounds active ingredient (a.i.)						
Corn	29.14	59.26	101.53	173.85	269.74	269.87	259.04	210.73	179.52	173.03	203.73
Soybeans	2.74	11.88	35.69	75.91	138.39	83.99	80.49	70.23	80.84	90.96	111.96
Potatoes	9.20	6.30	12.10	14.65	19.35	21.66	22.32	44.07	56.06	39.89	52.53
Cotton	76.54	77.27	97.05	76.62	49.19	34.10	37.90	85.88	87.16	63.06	37.56
Wheat	6.48	9.19	9.60	21.42	31.41	19.13	20.31	22.11	18.74	18.06	23.31
Sorghum	2.72	3.48	8.48	12.33	20.02	12.08	10.18	15.30	14.86	12.23	14.17
Peanuts	6.77	13.62	20.31	16.32	26.66	15.47	26.02	19.02	10.49	10.70	10.32
Rice	0.77	2.14	5.15	7.77	10.67	6.80	14.39	13.81	13.22	10.16	7.58
Tomatoes	9.88	8.82	9.17	5.31	5.94	5.93	13.94	18.02	14.30	15.30	9.70
Apples	9.17	6.31	2.85	2.39	5.34	4.27	8.80	8.10	10.05	7.96	7.28
Grapes	1.99	2.52	2.38	2.47	7.22	7.08	12.12	5.16	10.90	8.39	7.90
Subtotal	157.68	204.46	309.63	411.05	587.45	485.11	510.66	519.89	506.13	455.75	499.16
Other crops[1]	38.79	43.07	40.73	37.82	42.59	42.86	14.30	22.02	25.05	19.29	16.95
Total[2]	196.47	247.53	350.37	448.88	630.03	527.97	524.96	541.91	531.18	475.04	516.11
Herbicides	35.18	82.55	169.28	280.63	468.06	395.60	405.64	373.65	354.58	349.23	393.88
Insecticides	113.83	116.36	124.11	109.83	105.05	76.13	63.10	72.82	71.00	34.51	28.55
Fungicides	25.15	23.49	27.06	27.67	26.71	24.44	21.36	26.57	28.97	28.41	28.87
Other pesticides[2]	22.31	25.14	29.92	30.74	30.21	31.81	34.86	68.86	76.62	62.89	64.81
Total	196.47	247.53	350.37	448.88	630.03	527.97	524.96	541.91	531.18	475.04	516.11

[1]Other crops include: Barley, Grapefruit, Lemons, Lettuce, Peaches, Pears, Pecans, Sugarcane, and Sweetcorn.
[2]Other pesticides include soil fumigants, defoliants, desiccants, harvest aids, and plant growth regulators.

Sources: Economic Research Service with USDA and proprietary data. See Appendix 2.

Figure 2
Pesticide use in U.S. agriculture*—Comparing EPA estimates for total agriculture to ERS estimates for 21 crops and to planted acreage, 1960-2008

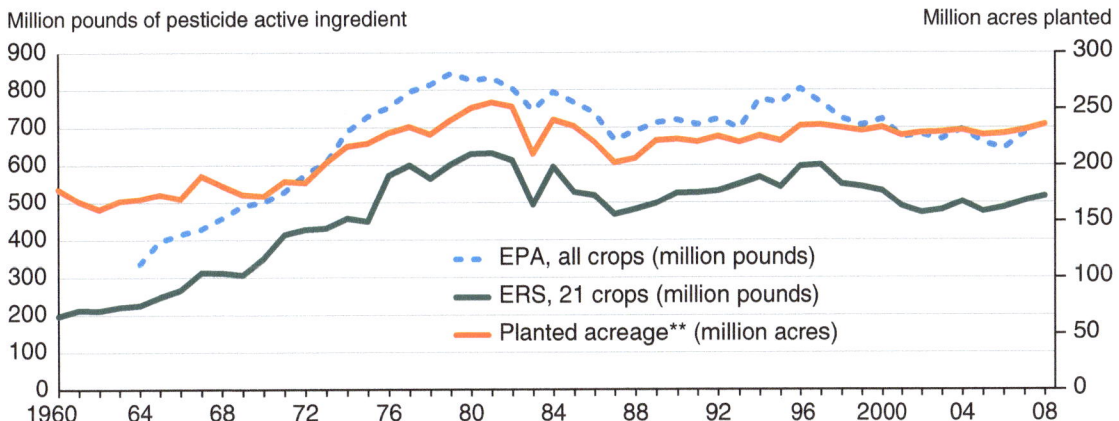

Million pounds of pesticide active ingredient

Million acres planted

- - - EPA, all crops (million pounds)
——— ERS, 21 crops (million pounds)
——— Planted acreage** (million acres)

*Conventional Pesticides. **Includes acreage of corn, cotton, soybean, wheat, and potatoes.
Sources: Economic Research Service with USDA and proprietary data (see appendix 2); EPA (1999, 2011).

Figure 3
Corn: acres treated with pesticides, 1952-2008

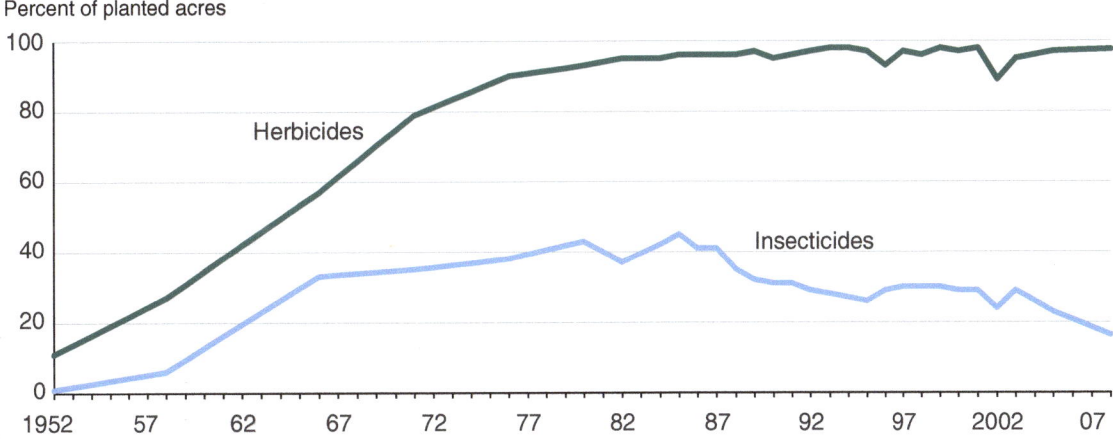

Percent of planted acres

Herbicides

Insecticides

Sources: Economic Research Service with USDA data. See Appendix 2.

over 90 percent of cotton and soybean acres were treated by 1980.[9] Thus, the diffusion of herbicides in U.S. agriculture followed the typical pattern of the diffusion of agricultural innovations (Griliches, 1957; Rogers, 1995).[10]

Increasing crop acreage also increased pesticide use. Total planted acreage of corn, cotton, potatoes, wheat, and, in particular, soybeans increased from the early 1960s to early 1980s, from 178

---

[9]Estimates of percent of acreage treated indicate how extensively herbicides, insecticides, fungicides, and other pesticides are used on a crop but not how intensively, since they do not reflect the number of treatments per acre, application rates, or the mixture of active ingredients applied.

[10]Diffusion curves are based on the notion that the current adoption rate is a function of the ultimate adoption level (ceiling) and the current adoption level. Adoption initially increases slowly as only the innovators (more venturesome and willing to assume risks) adopt. As information spreads, adoption rates increase. Finally, as the rate approaches the ceiling, the rate slows. At this point, most producers that find the innovation profitable have adopted. This process results in an S-shaped (sigmoid) diffusion curve (Fernandez-Cornejo et al., 2002).

Figure 4
Cotton: acres treated with pesticides, 1952-2008

Percent of planted acres

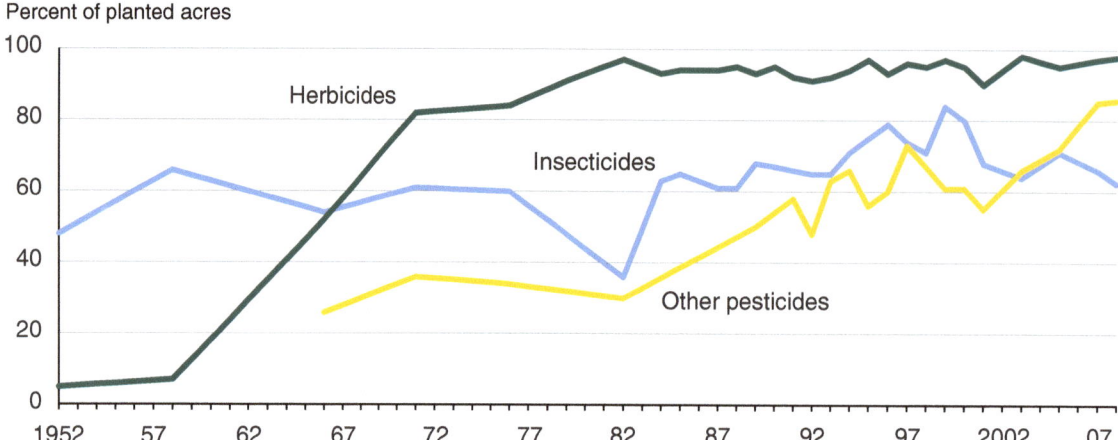

Sources: Economic Research Service with USDA data. See Appendix 2.

Figure 5
Soybeans: acres treated with pesticides, 1966-2006

Percent of acres treated

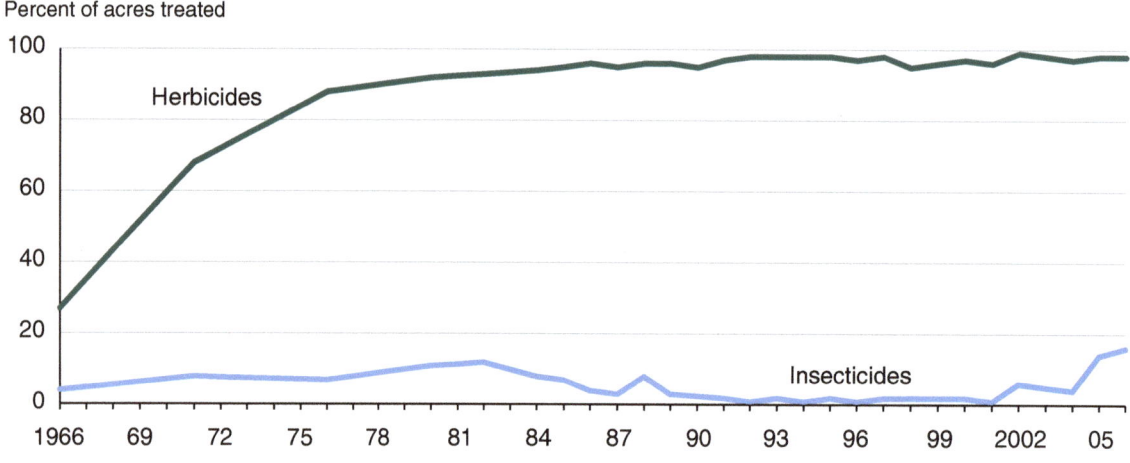

Sources: Economic Research Service with USDA data. See Appendix 2.

to 256 million acres. However, from 1960 to 1970, pesticide use increased faster than crop acreage, reflecting the effects of increasing share of crop acreage treated (fig. 2).[11, 12]

---

[11]Between 1960 and 1981, soybean acreage nearly tripled from 24 to 68 million acres, wheat acreage increased 60 percent from 55 to 88 million acres, cotton acreage decreased and potato acreage remained stable. While corn acreage was 81 million in 1960 and 84 million in 1981, it declined to the range of 65 to 70 million acres during the early 1960s, but increased during the 1970s to more than 80 million acres from 1976 to 1981, a period of rapid growth in corn insecticide and herbicide use.

[12]It could be argued that increased farm size increases pesticide use, but there is little economic research addressing this topic. Some economists have argued that risk-averse farmers increase pesticide use to reduce risk, and that increased farm size increases whole-farm risk aversion, encouraging more pesticide use (Osteen, 1987; Osteen et al., 1988; and Osteen and Fernandez-Cornejo, 2013). However, Osteen et al. (1988) showed that increased risk aversion could increase both application rates and the pest threshold for treatment. So, pesticide use could increase in some cases and decrease in others.

Figure 6
Wheat: acres treated with pesticides, 1952-2008

Percent of planted acres

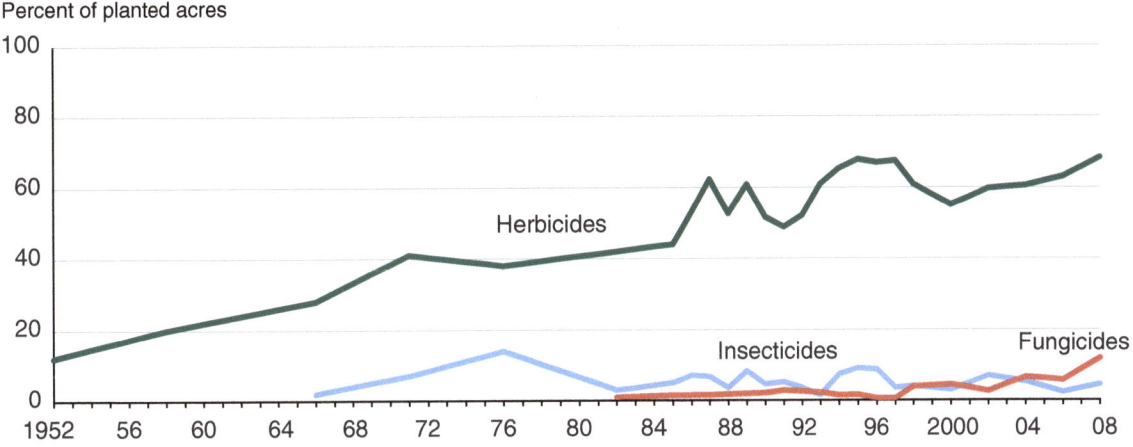

Sources: Economic Research Service with USDA data. See Appendix 2.

Figure 7
Potatoes: acres treated with pesticides, 1952-2008

Percent of planted acres

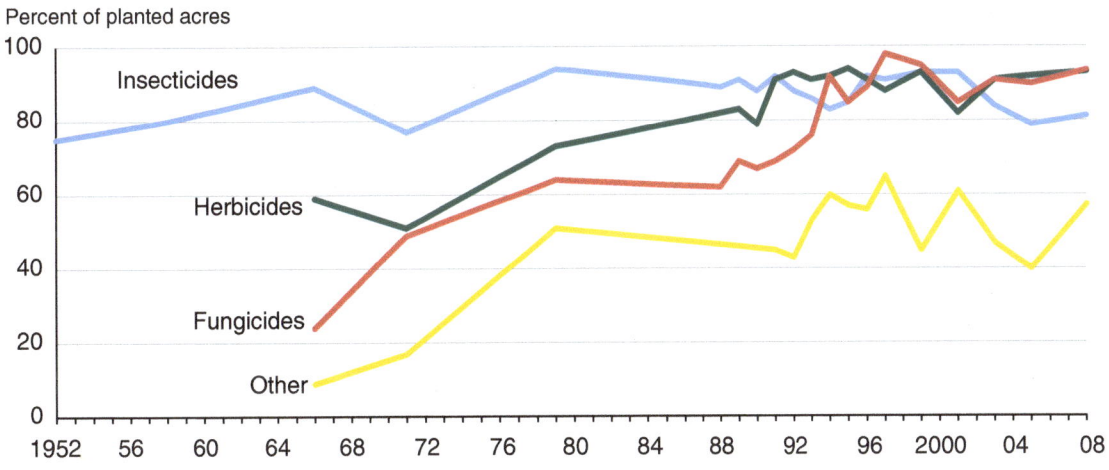

Sources: Economic Research Service with USDA data. See Appendix 2.

Herbicide Adoption Driven by Relative Prices

The NASS pesticide price index fell relative to the NASS wage, fuel, and crop indices from the late 1960s to about 1980, a period of rapid growth in pesticide use (see box, "Pesticide Prices and Relative Price Trends"). As several new effective products were introduced, the cost of pesticides fell relative to other pest control practices over the period and encouraged substitution of pesticides (particularly herbicides) for labor, fuel, and machinery use in pest control. In particular, the longrun decrease in herbicide prices relative to wage rates induced a strong labor-saving and herbicide-using bias in technological change in U.S. agriculture from 1960 to 1994. Fernandez-Cornejo and Pho (2002) found that a 1-percent decrease in the expected price ratio of herbicides to labor led in the long run to a 13.5-percent increase in the quantity ratio of herbicides to labor. Moreover, as Szmedra (1991) observes, the adoption of herbicides to control weeds substituted for labor and farm machinery. For example, the labor required to produce an acre of corn dropped from 13.2 hours in 1952 to 4.8 hours in 1976 while acreage treated with herbicides grew from 10 to 90 percent of corn planted acres (labor used per acre of corn production ranged between 2.1 and 3.5 hours in 1996 (Foreman, 2001)).

Stabilization of Herbicide Use and Decline in Insecticide Use Driven by Demand and Supply Factors

After 1980, the growth in aggregate herbicide use largely stopped because most corn, cotton, and soybean acreage was already being treated with herbicides. Still, herbicide use has continued to dominate the pesticide market; over the past three decades, more than 60 percent of the pounds of pesticides applied annually by U.S. farmers have been herbicides (76 percent in 2008, table 1). Together corn, cotton, soybeans, potatoes, and wheat accounted for 89 percent of herbicide use and 83 percent of total pesticide use in 2008 (fig. 8, appendix tables 3.1-3.5).

Changes in the planted acreages of these crops contributed to fluctuations in pesticide use from 1981 to 2008. Many of the high and low years in herbicide and pesticide use coincide with high and low years in total acreage of these crops (fig. 2).[13] However, declines in average application rates due to the introduction and use of new pesticides contributed to reductions in quantities used after 1981, while increased corn acreage and higher herbicide application rates account for increased pesticide use after 2002.

In addition to changes in the total quantity of herbicides applied, there have also been shifts in the herbicide active ingredients applied to major crops, as well as reductions in insecticide use. In 1968, atrazine and 2,4-D were among the top five pesticides used, but the other three were insecticides: toxaphene, DDT, and methyl parathion (fig. 9). In 2008, each of the top five herbicides (glyphosate, atrazine, acetochlor, metolachlor, and 2,4-D) were more heavily used than the top insecticide (chlorpyrifos) (fig. 10).[14] Figures 11 and 12 show the top herbicide active ingredients used in 1968 and

Figure 8

Pesticide use by crop, 21 selected crops, 2008, percent total pounds of active ingredient applied

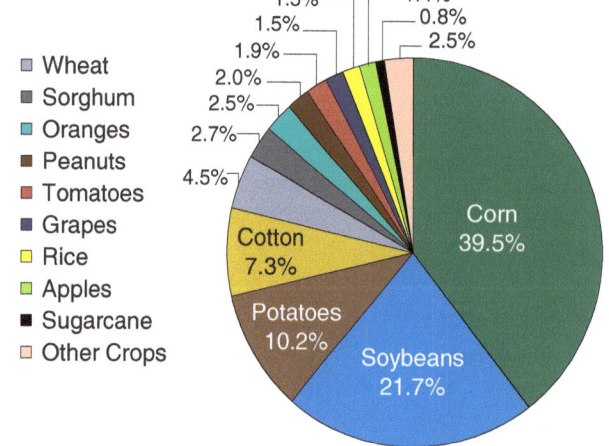

Note: "Other Crops" include: lettuce, pears, sweet corn, barley, peaches, grapefruit, pecans, and lemons.
Sources: Economic Research Service with USDA and proprietary data. See Appendix 2.

---

[13]Years when highs in fluctuations of pesticide use and combined corn, cotton, soybeans, wheat, and potato acreage coincide are 1981, 1984, 1994, 1997, 2004, and 2008, while years when lows coincide are 1983, 1987, 1995, and 2005.

[14]The fumigants metam-sodium and 1,3-dichloropropene (used in specialty crop production) are also among the top active ingredients used.

Pesticide Prices and Relative Price Trends

Relative price trends for crops, pesticides, and other inputs have influenced the cost effectiveness of pesticides and the amount used, given the comparative effects of different pesticides, non-pesticide practices, and management systems on pests and damages, which can also change over time. Overall, the NASS pesticide price index fell relative to the NASS wage and fuel indices from 1965 to 2008, while it increased relative to the NASS crop price index in some years and decreased in others, with essentially the same ratio between the pesticide and crop price indices in 1965 and 2007 (box figs. 3.1 and 3.2).

Box figure 3.1
Relative prices, 1965 - 2008: pesticides to wages, fuels, and crops

Index, 1965 = 1.0

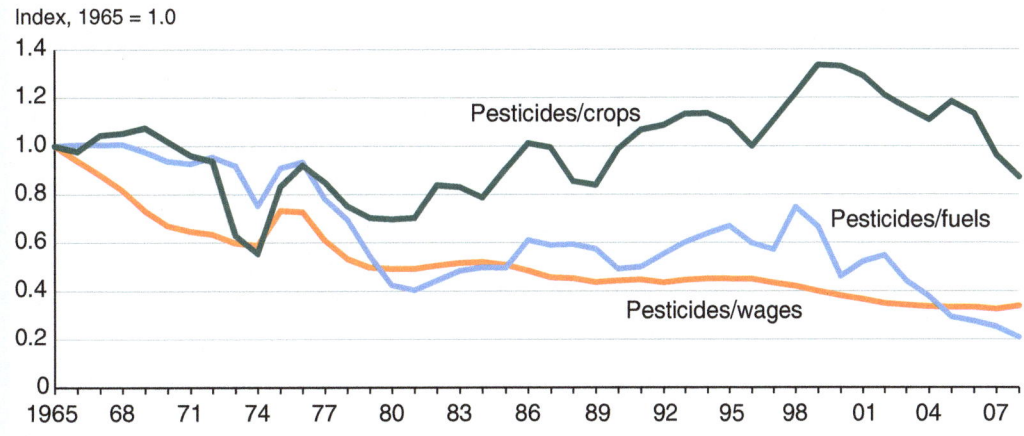

Source: USDA/NASS (*Agricultural Prices Summaries*).

Box figure 3.2
Price indices, 1965 - 2008: crops, wages, fuels, and pesticides

Index, 1965 = 1.0

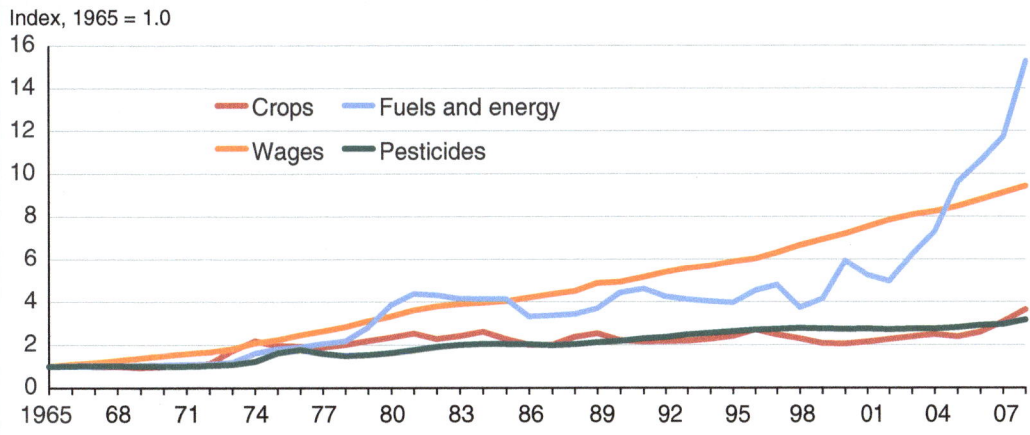

Source: USDA/NASS (*Agricultural Prices Summaries*).

Pesticide Use in U.S. Agriculture: 21 Selected Crops, 1960-2008, EIB-124
Economic Research Service/USDA

The pesticide price index fell relative to wage, fuel, and crop indices from the late 1960s to about 1980, a period of rapid growth in pesticide use. Relative price declines during that period helped reduce the cost of pesticides relative to other pest control practices and encouraged substitution of pesticides for labor, fuel, and machinery use in pest control. The increase in crop prices relative to pesticides (especially in 1973-74) also increased the returns to pesticides, encouraging greater use.

The pesticide price index rose relative to fuel and crop price indices from 1980 until the late 1990s, before falling in recent years. Increasing relative pesticide prices during this time period may have reflected high demand for pesticide use in crop production and contributed to use stabilizing after 1980. Since the late 1990s, the pesticide price index has declined relative to crops, wages, and fuels, reinforced by large increases in crop prices during 2005-08 and fuel prices during 2002-08, thus reverting to the longer term trend, encouraging substitution of pesticides for labor, fuel, and machinery used in pest control and more pesticide use to protect crop values.[1]

Since 1990, NASS insecticide and fungicide price indices have risen more rapidly than the herbicide price index (box fig. 3.3). Nonetheless, because herbicides, particularly atrazine and glyphosate, dominate the pesticide market, we can identify underlying trends in pesticide prices by tracking major herbicide prices (box table 3.1). Among major agricultural herbicides, prices have risen sharply only for metolachlor; the price of atrazine grew only 1 percent per year and the price of glyphosate actually fell. In fact, in real terms, herbicide prices have fallen since 1990.

Box figure 3.3
Agricultural input price indices, 1990-2008

Index, 1991=100

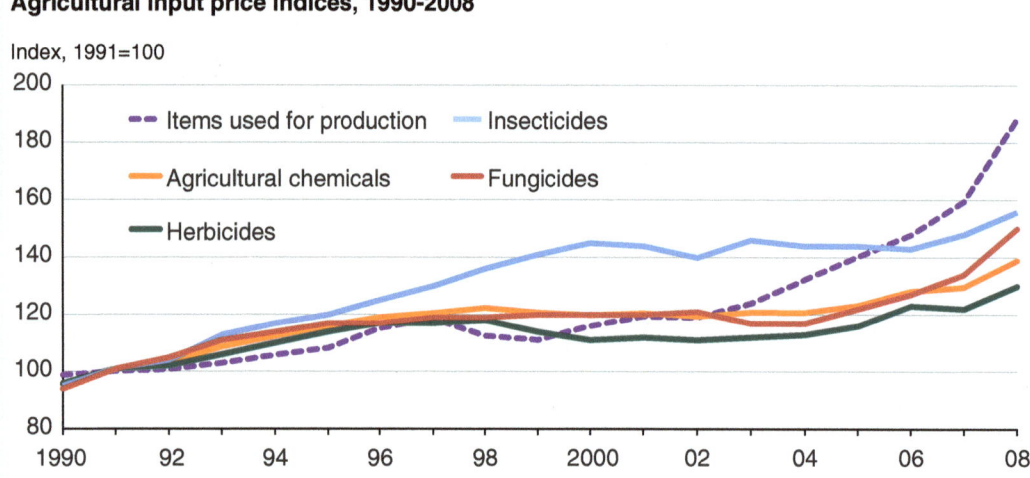

Source: USDA/NASS (*Agricultural Prices Summaries*).

---

[1]Since 1990, pesticide prices have risen less rapidly than the prices of all agricultural inputs. While the NASS index for agricultural input prices ("Items Used for Production") rose at an average annual rate of 3.7 percent from 1990 to 2008, the NASS pesticide price index rose by only 1.7 percent per year. Recently, high fertilizer and fuel costs have accelerated the growth in the "all inputs" price index compared to the pesticide price index. From 2000 to 2008, the NASS index for agricultural input prices grew at an average rate of 6.8 percent per year. The pesticide price index rose by only 1.9 percent per year.

By contrast, prices of some insecticides and fungicides—such as carbaryl, propargite, and captan—showed robust growth.

Box table 3.1
Selected pesticide prices, 1990-2008

Active ingredient	1990	1995	2000	2001	2002	2003	2004	2005	2006	2007	2008
						Dollars					
Selected herbicides (gallon)											
Atrazine (Aatrex) 4#/Gal L	NA	14.4	13.6	12.5	12.2	12.3	12.2	12.4	12.1	12.2	15.3
Glyphosate (Roundup) 4#/Gal EC	NA	54.1	43.3	44.5	43.5	43.3	39.7	33.8	29.3	28.9	40.5
Metolachlor (Dual) 8#/Gal EC	55.5	67.7	82.6	94.5	99	104	106	108	107	NA	NA
Selected insecticides (pound)											
Carbaryl (Sevin) 80% S, SP, or WP	3.5	4.6	5.4	5.8	5.4	5.5	5.9	5.9	5.5	6.4	7.1
Propargite (Comite) 30% WP	NA	5.9	6.9	6.1	6.3	6.6	6.4	7	7.5	8.7	9.2
Selected fungicides (pound)											
Captan 50% WP	2	3.3	3.5	3.6	3.8	3.5	3.5	3.7	3.9	4.6	5.5

Formulations: EC - Emulsifiable Concentrate, G - Granular, L - Liquid, S- Solution, SP - Soluble Powder, WP - Wettable Powder. NA - Not avilable.

Source: USDA/ NASS (Agricultural Price Summaries).

2008. Atrazine, 2,4-D, and trifluralin maintain their top 10 ranking in both years. Some herbicides that were ranked in the top 10 in 2008, such as glyphosate and metolachlor were not available in 1968. Other herbicides ranked among the top 10 in 1968 were not so ranked in 2008; for example dinoseb and vernolate, which were no longer used.

Herbicide-Tolerant (HT) Crops Transform Herbicide Mix

Several factors appear to have driven changes in herbicide use: crop and input price changes, the introduction of new herbicides, adoption of herbicide-tolerant (HT) crops, a shift toward conservation tillage systems, pesticide regulation, and government policies such as the incentives for ethanol producers.

In particular, the introduction of HT crops in the mid-1990s augmented pest management options by permitting the use of more effective and less toxic herbicides (such as glyphosate or glufosinate) than previous ones that would have destroyed the crops as well as the weeds. Cotton, soybean, and

Pesticide Use in U.S. Agriculture: 21 Selected Crops, 1960-2008, EIB-124
Economic Research Service/USDA

Figure 9
Pesticide use by active ingredient (a.i.), 21 selected crops in 1968, percent total pounds of a.i. applied[1]

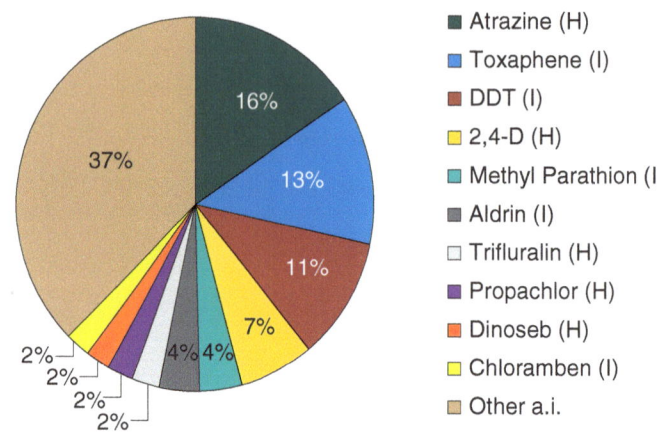

- ■ Atrazine (H)
- ■ Toxaphene (I)
- ■ DDT (I)
- ■ 2,4-D (H)
- ■ Methyl Parathion (I)
- ■ Aldrin (I)
- □ Trifluralin (H)
- ■ Propachlor (H)
- ■ Dinoseb (H)
- ■ Chloramben (I)
- ■ Other a.i.

[1]This graph shows the top pesticide a.i. (herbicide = H, insecticide = I) used in 1968.
Sources: Economic Research Service with USDA and proprietary data. See Appendix 2.

Figure 10
Pesticide use by active ingredient (a.i.), 21 selected crops in 2008, percent total pounds of a.i. applied[1]

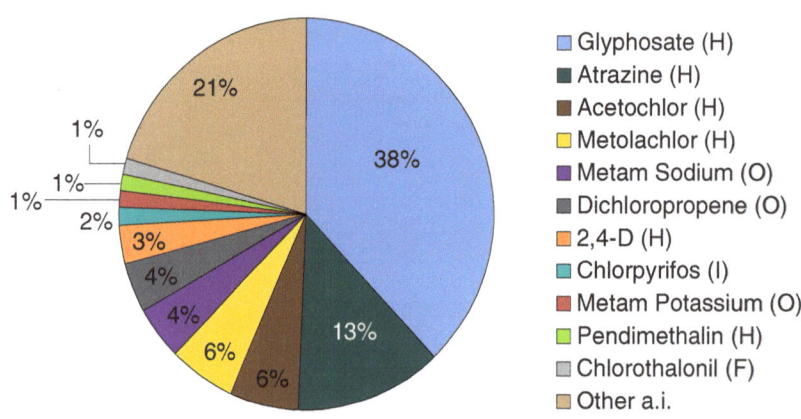

- ■ Glyphosate (H)
- ■ Atrazine (H)
- ■ Acetochlor (H)
- □ Metolachlor (H)
- ■ Metam Sodium (O)
- ■ Dichloropropene (O)
- ■ 2,4-D (H)
- ■ Chlorpyrifos (I)
- ■ Metam Potassium (O)
- ■ Pendimethalin (H)
- □ Chlorothalonil (F)
- ■ Other a.i.

[1]This graph shows the top pesticide a.i. (herbicide = H, insecticide = I, fungicide = F, and other = O) used in 2008.
Sources: Economic Research Service with USDA and proprietary data. See Appendix 2.

corn varieties designed to resist glyphosate, a broad-spectrum herbicide, were first marketed in 1996. Pounds of glyphosate per planted acre of soybeans, corn, and cotton rose in almost every year since 1996 while pounds of all other herbicides (per acre) fell (figs. 13-15; appendix tables 3.1-3.3).

Due to the substantial benefits provided to farmers (Carpenter and Gianessi, 1999; Fernandez-Cornejo and Caswell, 2006), HT seed adoption was most rapid and widespread among U.S. soybean farmers. By 2008, over 90 percent of soybean acres were planted with HT seeds (fig. 14). HT soybean production sharply boosted glyphosate use on soybeans from 0.17 pound per planted acre (a total of 11 million pounds applied) in 1996 to 1.26 pounds per planted acre (95 million pounds) in 2008. Pounds of all other herbicides applied to soybeans declined considerably from 1.02 pounds per planted acre in 1996 to 0.14 pound in 2008.

HT corn adoption increased from 3 percent of planted acres in 1996 to just over 60 percent of planted acres in 2008 (fig. 13). Consequently, glyphosate use jumped from 0.04 pound per planted

Figure 11
Herbicide use by active ingredient (a.i.), 21 selected crops in 1968, percent total pounds a.i. applied[1]

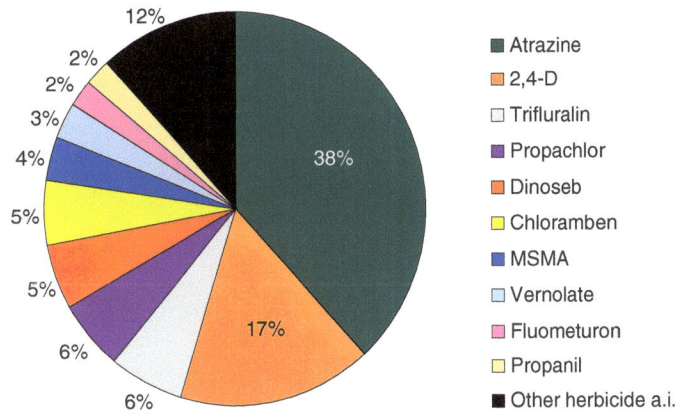

¹This graph shows the top herbicide a.i. used in 1968.
Sources: Economic Research Service with USDA and proprietary data. See Appendix 2.

Figure 12
Herbicide use by active ingredient (a.i.), 21 selected crops in 2008, percent total a.i. applied[1]

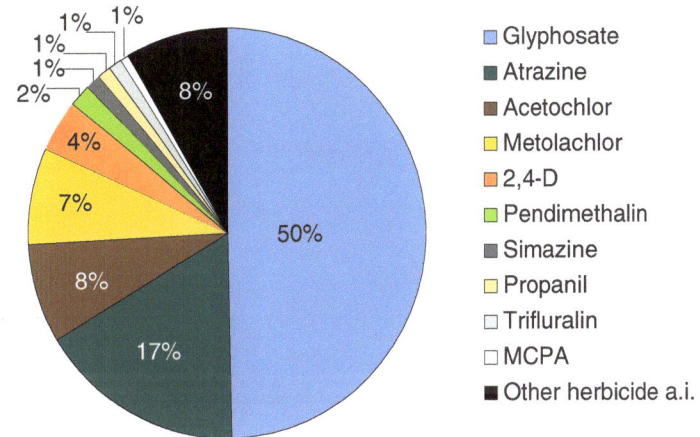

¹This graph shows the top herbicide a.i. used in 2008.
Sources: Economic Research Service with USDA and proprietary data. See Appendix 2.

acre (a total of 3 million pounds) in 1996 to 0.79 pound per acre (68 million pounds) in 2008. Glyphosate is now the single most heavily used corn herbicide. Other herbicides applied to corn fell from about 2.6 pounds per planted acre in 1996 to 1.5 pounds in 2008.

By 2008, about 70 percent of cotton acreage was planted with HT seed. Correspondingly, glyphosate use increased from about 0.09 pound per planted acre (1.3 million pounds) in 1996 to 1.45 pounds (14 million pounds) in 2008 (fig. 15). Other herbicides applied to cotton fell from about 2 pounds per acre in 1996 to 0.92 pounds per acre in 2008.

While the adoption of HT crops has led to the substitution of the more environmentally benign herbicide, glyphosate, for other herbicides (NRC, 2010),[15] overreliance on a limited number of herbi-

---

[15]Atrazine, an herbicide widely used during the last half century despite environmental concerns, has remained the second most applied herbicide since 2000 even though its share of herbicide use has declined (fig. 11).

Pesticide Use in U.S. Agriculture: 21 Selected Crops, 1960-2008, EIB-124
Economic Research Service/USDA

Figure 13
Pounds of herbicide active ingredient (a.i.) per planted acre and percent acres of herbicide-tolerant corn, 1996-2008

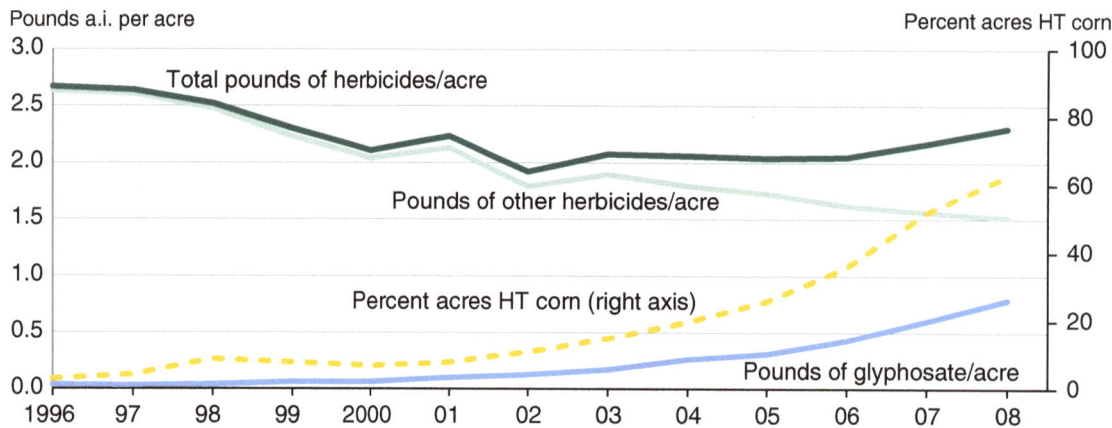

Sources: Pesticides: Economic Research Service with USDA and proprietary data. See Appendix 2; HT corn: Fernandez-Cornejo (2012).

Figure 14
Pounds of herbicide active ingredient (a.i.) per planted acre and percent acres of herbicide-tolerant soybeans, 1996-2008

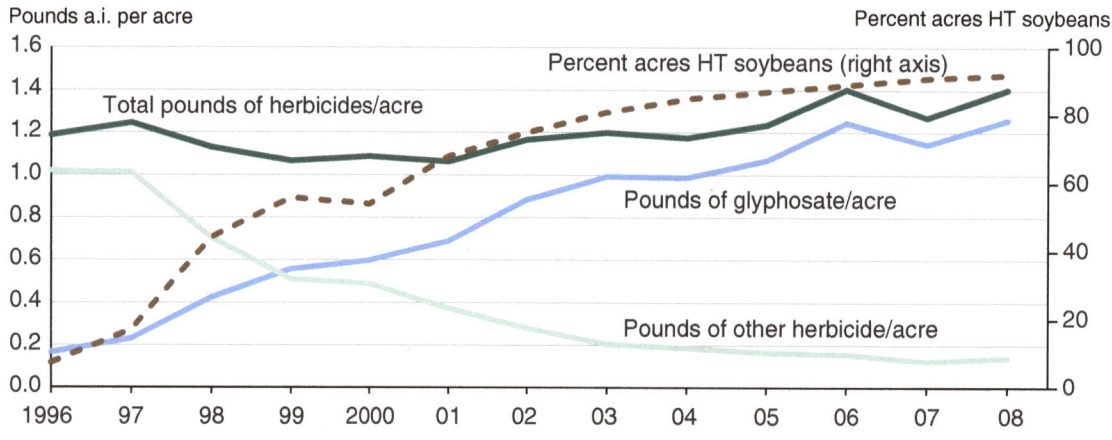

Sources: Pesticides: Economic Research Service with USDA and proprietary data. See Appendix 2; HT soybeans: Fernandez-Cornejo (2012).

cides may accelerate weed resistance to those chemicals. Resistance management strategies include weed scouting, rotating between crops treated with different herbicides and using weed management practices, rotating glyphosate with herbicides that have a different mode of action, limiting glyphosate applications to two over a 2-year period, or using cultivation or other mechanical weed control practices (Boerboom and Owen, 2006). Farmers can also manage resistance by using multiple herbicides—atrazine, s-metolachlor or acetochlor, and/or mesotrione with glyphosate on corn, for example (Loux et al., 2013).

Insecticide Use Declines as Herbicides Grow

Insecticide use by U.S. farmers has fallen as herbicide use has grown. Insecticide use was much higher in the 1960s and 1970s than in later years. It peaked at 158 million pounds in 1972, and has declined most years thereafter, ending at 29 million pounds in 2008 (fig. 1). In the 1950s, insecticides were widely

Figure 15

Pounds of herbicide active ingredient (a.i.) per planted acre and percent acres of herbicide tolerant cotton, 1996-2008

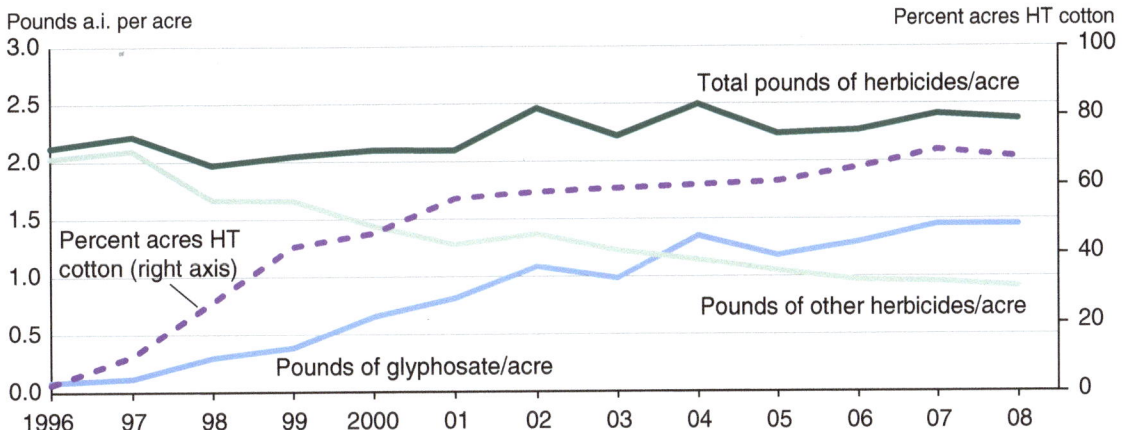

Sources: Pesticides: Economic Research Service with USDA and proprietary data. See Appendix 2; HT cotton: Fernandez-Cornejo (2012).

used on a variety of crops: cotton, tobacco, fruits, potatoes, and other vegetables. However, insecticides were applied to less than 10 percent of corn acreage during the 1950s. This share increased rapidly to 38 percent by 1976, reached 45 percent in the mid-1980s, but fell to about 16 percent in 2008 (fig. 3).

The proportion of cotton acreage treated with insecticides varied between about 50 and 70 percent from the 1950s to mid-1990s, exceeding 80 percent in 1999–2000 before declining to about 60 percent in 2008 (interpolated) (fig. 4). In most years, insecticides were used on less than 10 percent of soybean and wheat acres (figs. 5 and 6). Historically, the proportion of many vegetable and fruit acres treated with insecticides has been high. For example, the share of treated potato acreage exceeded 75 percent since the 1950s, exceeded 90 percent most years between 1978 and 2001 but declined to around 80 percent in 2008 (fig. 7).

Insecticide use and active ingredients applied have fluctuated with changes in crop acreage, changes in pest pressure, agricultural practices, pesticide regulation, technology, and other factors. For example, the Federal ban on some organochlorines, such as DDT, forced growers to find new pest control solutions and pesticide manufacturers to develop new crop protection products in the 1970s. Also, higher pest pressure in some years resulted in higher rates of insecticide application. As some older insecticides became less effective due to pest resistance, farmers applied at higher rates and/ or used new insecticides. DDT and toxaphene (used primarily in cotton production) dominated insecticide use in 1968 (fig. 16), but were displaced by other materials. Chloropyrifos and aldicarb became especially important in recent years (fig. 17).[16] Newer insecticides applied at low rates, such as synthetic pyrethroids (e.g. permethin and cypermethrin) and neo-nicotinoids (e.g., imidicloprid and clothianidin), have become widely used. (Osteen and Fernandez-Cornejo (2013) provide a more detailed discussion of changing insecticide use over time.) Some insecticides are now applied as seed treatments and are not generally captured by pesticide use surveys.

In 1996, genetically engineered insect-resistant corn and cotton varieties (Bt corn and cotton) were introduced commercially. These Bt crops carry a gene of the soil bacterium *Bacillus thuringiensis* (Bt), which induces plants to produce a protein that is harmful to some insects, including the

---

[16]In August 2010, EPA initiated action to terminate uses of aldicarb. It is being phased out with all uses scheduled to end in August 2018. http://www.epa.gov/oppsrrd1/REDs/factsheets/aldicarb_fs.html

Figure 16
Insecticide use by active ingredient (a.i.), 21 selected crops in 1968, percent total pounds a.i. applied[1]

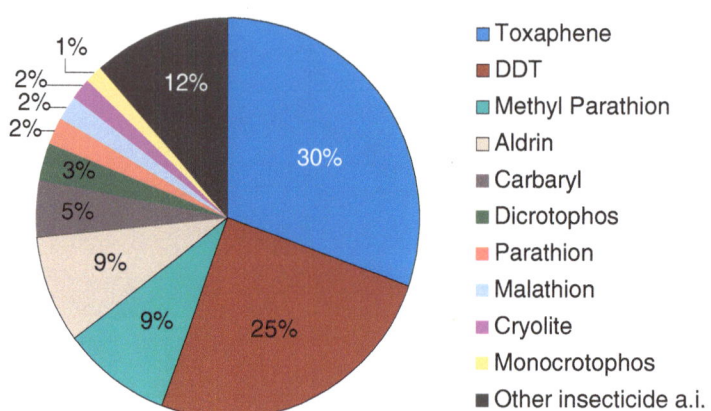

- Toxaphene
- DDT
- Methyl Parathion
- Aldrin
- Carbaryl
- Dicrotophos
- Parathion
- Malathion
- Cryolite
- Monocrotophos
- Other insecticide a.i.

[1]This graph shows the top pesticide a.i. used in 1968.
Sources: Economic Research Service with USDA and proprietary data. See Appendix 2.

Figure 17
Insecticide use by active ingredient (a.i.), 21 selected crops in 2008, percent total pounds a.i. applied[1]

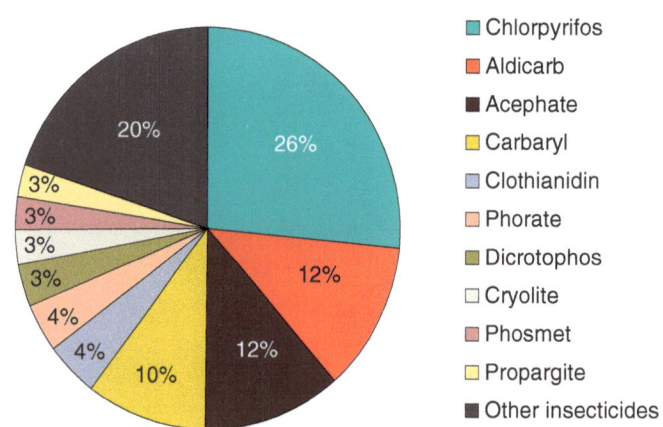

- Chlorpyrifos
- Aldicarb
- Acephate
- Carbaryl
- Clothianidin
- Phorate
- Dicrotophos
- Cryolite
- Phosmet
- Propargite
- Other insecticides

[1]This graph shows the top pesticide a.i. used in 2008.
Sources: Economic Research Service with USDA and proprietary data. See Appendix 2.

European corn borer and corn rootworm. Adoption of Bt corn and cotton has been associated with a reduction in insecticide use (Fernandez-Cornejo and Caswell, 2006; NRC, 2010). For example, corn farmers using seeds without Bt traits applied 0.1 pound of insecticide per planted acre in 2001 and 0.09 pound of insecticide in 2005. By contrast, corn farmers using Bt seeds applied 0.07 pound per planted acre in 2001 and 0.05 pound in 2005 (Fernandez-Cornejo and Wechsler, 2012).[17]

---

[17]Research by ERS and others suggests that, controlling for other factors, insecticide use declined with the adoption of Bt corn and Bt cotton (Fernandez-Cornejo and Caswell, 2004). It should be noted, however, that by protecting the plant from certain pests, Bt crops can also prevent yield losses compared with non-GE hybrids, particularly when pest infestation is high. This effect is particularly important for Bt corn, which was introduced in the mid-1990s to control the European corn borer. Since chemical control of the European corn borer was not always profitable, and timely application was difficult, many farmers accepted yield losses rather than incur the expense and uncertainty of chemical control. For those farmers, the introduction of Bt corn resulted in yield gains rather than pesticide savings. On the other hand, another type of Bt corn introduced in 2003 to provide resistance against the corn rootworm, which was previously controlled using chemical insecticides, does provide substantial insecticide savings (Fernandez-Cornejo and Caswell, 2006).

The rapid adoption of Bt seed since 1996 may have contributed to declining shares of corn and cotton acreage treated with insecticides. Insecticide use for corn production, which had peaked in the late 1970s and 1980s at 0.35-0.45 pound per acre, declined throughout the 1990s and 2000s to under 0.05 pound per planted acre in recent years (fig. 18, appendix table 4.1).[18] Insecticide use for cotton, which had peaked at 9.5 pounds per planted acre in 1967, trended downward to between 1 and 2 pounds per planted acre in the 1980s (as lower dose insecticides replaced older, higher dose insecticides). Since the adoption of Bt cotton, insecticide use in cotton has declined to less than 1 pound per planted acre (fig. 19, appendix table 3.3). Higher insecticide application rates on cotton in 1999 and 2000 can be attributed to the boll weevil eradication program (see "Cotton Pesticide Trends" in Appendix 4).

Figure 18
Pounds of insecticide active ingredient (a.i.) per planted acre and percent acres of Bt corn, 1996-2008

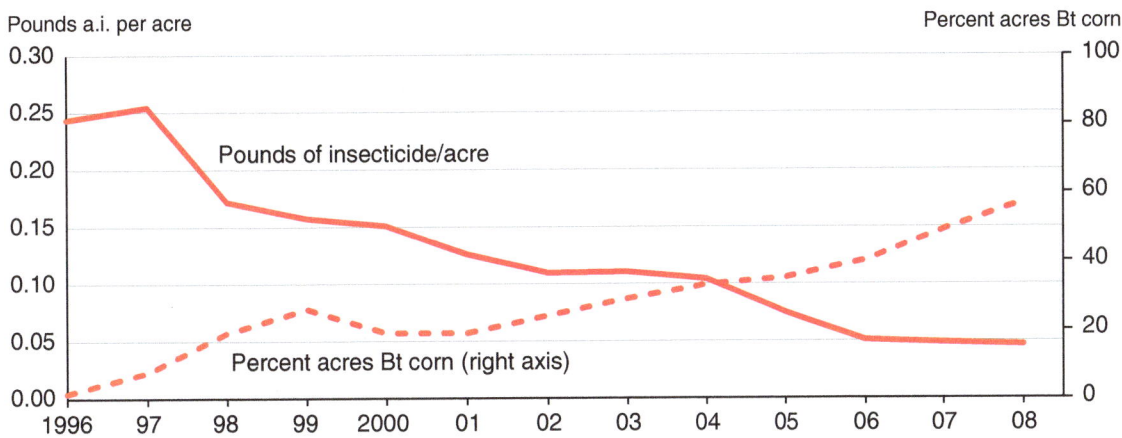

Sources: Pesticides: Economic Research Service with USDA and proprietary data. See Appendix 2. Bt corn: Fernandez-Cornejo (2012).

Figure 19
Pounds of insecticide active ingredient (a.i.) per planted acre and percent acres of Bt cotton, 1996-2008

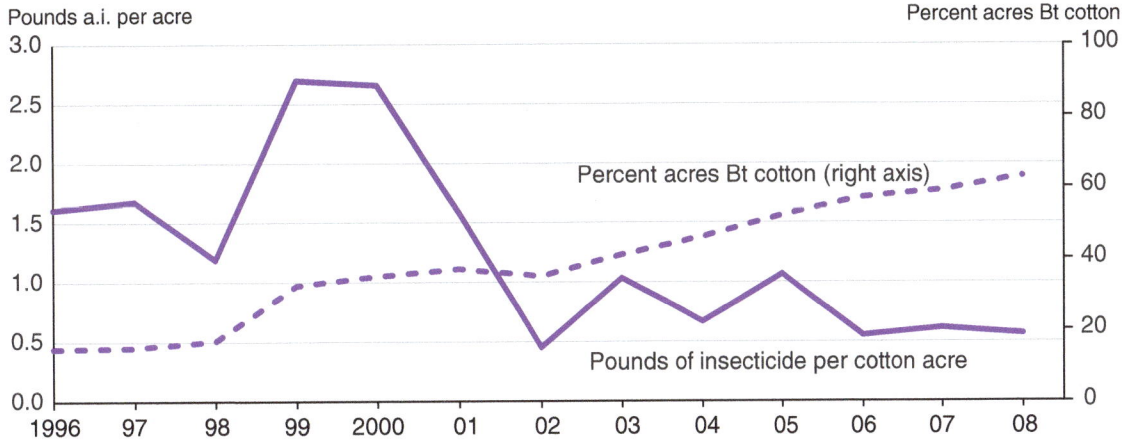

Sources: Pesticides: Economic Research Service with USDA and proprietary data. See Appendix 2. Bt cotton: Fernandez-Cornejo (2012).

---

[18]In terms of total pounds of insecticide active ingredient, corn producers reduced their use from 25-35 million pounds in the 1980s to about 19 million in 1995 and 4 million pounds in 2006-2008 (appendix table 4.1). Cotton producers reduced their use from their 1967 peak of 90 million pounds to 28 million pounds in 1995 and 5-8 million pounds in 2006-2008 (appendix table 4.3).

Fungicides Used Mainly on Fruits and Vegetables

Fungicides' share of pesticide use has remained at 7 percent or less since 1971, down from 11-13 percent in the early 1960s (fig. 1). Excluding seed treatments, fungicides have generally been applied to a greater percentage of fruit and vegetable acres than to corn, soybean, cotton, and wheat acres, but treatments on corn, soybeans, and wheat have increased in recent years. For example, treated potato acres increased from 24 percent in 1966 to 85-95 percent in the 1990s onward (fig. 7). By contrast, the acreage of corn, cotton, soybean, and wheat treated with fungicides is generally less than 10 percent (excluding seed treatments).

Among the most prominent fungicides used in 2008 were chlorothalonil (accounting for 24 percent of active ingredient pounds), copper compounds (17 percent), mancozeb (15 percent), captan (6 percent), maneb (3 percent), and ziram (3 percent), all of which have been used for many years (fig. 20). However, in recent years, corn, soybean, and wheat growers have applied such materials as pyraclostrobin (that accounted for 7 percent of all pounds a.i. in 2008), propiconazole (5 percent), azoxystrobin (4 percent), and tebuconazole (2 percent). In 2006 and 2007, EPA granted emergency exemptions for fungicides to control soybean rust (such as, cyroconazole, famoxadone, flusilazole, metconazole, and prothioconazole).

Cotton and Potatoes Are Major Users of Other Pesticides

Other pesticides—which include soil fumigants, defoliants, desiccants, harvest aids, and plant growth regulators—generally accounted for about 11 percent or less of total pesticide use from 1960 to 1992, but increased to 17 percent in 2002, before declining to 13 percent in 2008 (fig. 1). The total quantity increased from 22 million pounds in 1960 to 80 million pounds in 2002, declining to 65 million pounds in 2008. Despite their declining share of total use, the quantity of other pesticides applied to the 21 crops we examined has exceeded that of insecticides since the mid-1990s. These chemicals are commonly applied to cotton, potatoes, other vegetables, and many fruit crops, but cotton and potatoes accounted for 82 percent of other pesticide use in 2008 (appendix tables 4.3 and 4.4).

Figure 20

Fungicide use by active ingredient (a.i.), 21 selected crops in 2008, percent total pounds a.i. applied[1]

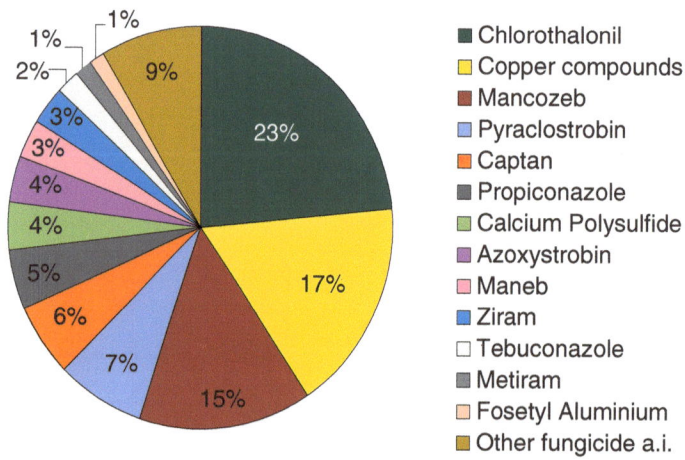

[1]Calcium polysulfide can be used as a fungicide, miticide, or insecticide.
Sources: Economic Research Service with USDA and proprietary data. See Appendix 2.

Cotton is a major user of plant growth regulators and harvest aids. Cotton acreage treated with other pesticides increased from 26 percent in 1966 to 70 percent in the late 1990s and about 85 percent in 2007 (fig. 4). The share of potato acreage treated, including fumigants and harvest aids, increased from 9 percent in 1966 to over 50 percent in the late 1970s, and fluctuated between 40 and 70 percent since 1998 (fig. 7). The increased use of fumigants on potatoes and other vegetables such as tomatoes, contributed to the increase in the quantity of other pesticide use through 2000. However, methyl bromide use on tomatoes and other vegetables declined since 1994 due to the Montreal Protocol phase-out.

Five Major Crops Account for Four-Fifths of Pesticide Use

In 2008, corn, soybeans, cotton, wheat, and potatoes accounted for about 80 percent of the pesticide quantity applied to the 21 crops we examined (fig. 8). The share of pesticides applied to corn production reached a peak in 1985 (over half of the pesticides applied to all 21 crops), but has dropped in recent years (table 2). Corn continues to receive the highest share of pesticide use, about 39 percent of the pounds applied in 2008. Twenty-two percent of the total pesticide pounds applied were devoted to soybean production. Potatoes' share rose significantly in the 1990s and reached about 10 percent by 2008. Meanwhile, the pounds of pesticide applied to cotton has trended downward due to the replacement of DDT and other older insecticides with more effective products, the eradication of the boll weevil, and the adoption of Bt cotton. Cotton accounted for just over 7 percent of the total pesticide pounds applied while wheat accounted for less than 5 percent in 2008.

Appendix 4 discusses these trends for the top five crops in more detail.

Pesticide Quality

Inherent differences in pesticide characteristics or quality complicate the direct comparison of observed prices and quantities of pesticides over time. This comparison can be facilitated by using price and quantity indices adjusted for quality. Quality-adjusted price and quantity indices have proven to be useful for products subject to rapid technological change such as pesticides. New and better pesticide active ingredients (more effective and less harmful to human health and the environment) have frequently been introduced while other active ingredients have been banned or voluntarily canceled by their manufacturers (Fernandez-Cornejo and Jans, 1995). As a result, there are several hundred pesticide active ingredients in use.[19] (See box, "Quality Adjustment of Price and Quantity Indices for Pesticides," which illustrates the estimation of the quality-adjusted price and quantity indices for pesticide applied to four major crops: corn, cotton, soybeans, and sorghum over 1968-2008.) Appendix table 2.2 shows a list of the active ingredients considered in this study.

Because of quality improvements (improved pest control effectiveness or lower application rates), adjusted pesticide prices (constant quality) increase less than the unadjusted prices (unadjusted or actual prices reflect technological improvements and therefore are higher). Similarly, quality-adjusted quantities are higher than unadjusted quantities because farmers would have had to use more pesticides if pesticide quality had remained constant instead of improving (Fernandez-Cornejo and Jans, 1995).

---

[19]Per EPA, there were 684 active ingredients registered for agricultural use in 2007
http://www.epa.gov/pesticides/pestsales/07pestsales/usage2007_2.htm

Quality Adjustment of Pesticides

As an illustration, the quality-adjusted prices were calculated using a hedonic approach and represent the prices that would have been obtained if quality had remained constant. The adjustment process uses a hedonic function, which entails expressing the price of pesticides as a function of their quality characteristics (Fernandez-Cornejo and Jans, 1995). After controlling for observable characteristics, it is possible to estimate quality-adjusted price indices. Quality-adjusted quantity indices are computed by dividing pesticide expenditures by the quality-adjusted price indices.

The quality characteristics considered in this illustration are pesticide potency, hazardous characteristics, and persistence. Pesticide potency is inversely related to the application rate per crop year, which can be viewed as the dosage and is equal to the pounds of active ingredient applied per acre in one application multiplied by the number of applications made in a year (Fernandez-Cornejo and Jans, 1995). Thus, the lower the rate needed to achieve a degree of pest control, the more potent the pesticide is. Hazardous characteristics are measured by chronic toxicity scores, and persistence is measured by the pesticide's half-life. The chronic toxicity index is the inverse of the water quality threshold (which measures the concentration in parts per billion) and serves as the environmental-risk indicator for humans from drinking water (Kellogg et al., 2002). The lower the index, the lower is the potential environmental risk for the chemical. The persistence indicator is defined by the share of pesticides with a half-life less than 60 days (Fernandez-Cornejo and Jans, 1995). The lower the indicator, the less persistent the pesticide is.

To illustrate, box figure 4.1 provides an estimate of the (weighted) average quality measures for the pesticides used in four major crops. Three tendencies can be noted. First, pesticide *potency* has increased in the 1970s and 2000s as the application rate per crop year needed to achieve a degree of pest control declined. This is due in part to the use of improved pesticides and GE seed. Second, average chronic *toxicity* declined, as toxic products applied to cotton (such as DDT and toxaphene) and to corn (such as aldrin) were banned (particularly in the 1970s and early 1980s). Other factors affecting toxicity were the use of less toxic insecticides, such as carbaryl and chlorpyrifos, the introduction of pyrethroids, the use of malathion in the boll weevil eradication program, and the use of Bt cotton since 1996. Third, *persistence* fell during the 1970s after the bans of DDT and aldrin, then increased during the 1980s and early 1990s (in part with the use of high-persistence products such as metolachlor and pendimethalin); persistence has declined in recent years, reflecting the rapid increase in glyphosate use. Glyphosate has low chronic toxicity (a high chronic score) and relatively low persistence relative to the herbicides that it has replaced. As the NRC (2010) report states, glyphosate "is biodegraded by soil bacteria and it has a very low toxicity to mammals, birds, and fish (Malik et al., 1989)." Also, some newer insecticides such as clothianidin, and herbicides, such as mesotrione and nicosulfuron in corn production, are applied at low rates and have low chronic toxicity and low persistence.

Box figure 4.1

Average quality characteristics of pesticides applied to four major crops, 1968-2008

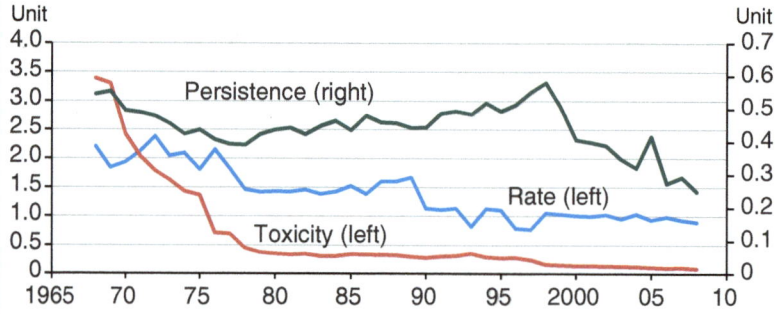

Rate: Pounds of active ingredient applied per acre in one application times the number of applications per year.

Sources: Estimates based on USDA and proprietary data (Appendix 2) for four major crops: corn, soybeans, cotton and sorghum.

Unadjusted (actual) pesticide prices stopped rising in the mid-1990s despite quality improvements (fig. 21), and the decline in glyphosate price associated with its patent expiration in 2000 may have been a factor. Glyphosate replaced other herbicides, as adoption of herbicide-tolerant (HT) crops and conservation tillage increased. Quality-adjusted prices have declined since the mid-1990s because unadjusted prices stagnated and the quality of pesticides improved. In particular, glyphosate contributed to quality improvement because it is more effective and less toxic than many of the conventional herbicides that it replaced (NRC, 2010).

While unadjusted pesticide quantities show little upward movement after 1997 (fig. 22), the quantity indices adjusted for quality (i.e., at constant quality) show a substantial increase. This implies that if pesticide quality had not improved (e.g., if herbicides had not become more effective), total pesticide use would have increased rapidly from 1996 to 2008.

Figure 21
Quality adjusted pesticide prices applied to 4 major crops, 1968-2008

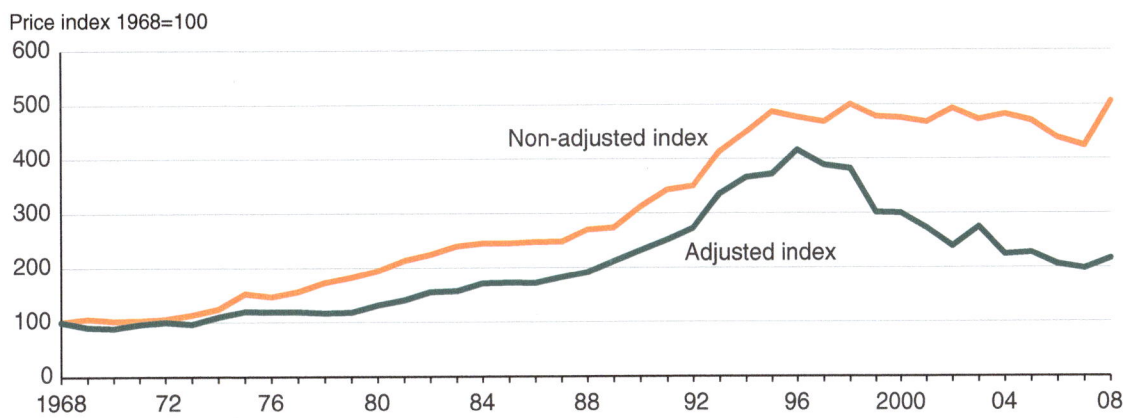

Note: Quality adjusted indices were calculated for pesticides applied to four major crops: corn, soybeans, cotton and sorghum.
Source: Estimates based on USDA and proprietary data (Appendix 2).

Figure 22
Quality adjusted pesticide quantities applied to 4 major crops, 1968-2008

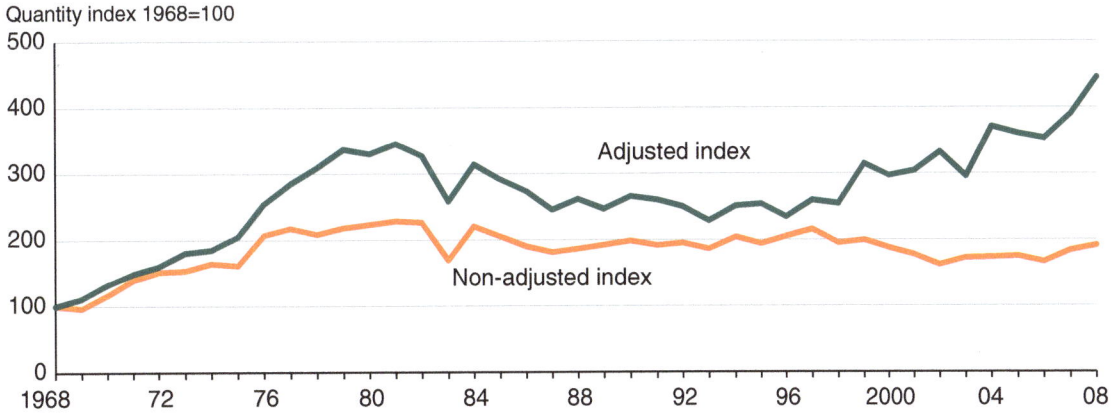

Note: Quality adjusted indices were calculated for pesticides applied to four major crops: corn, soybeans, cotton and sorghum.
Source: Estimates based on USDA and proprietary data (Appendix 2).

Pesticide Use in U.S. Agriculture: 21 Selected Crops, 1960-2008, EIB-124
Economic Research Service/USDA

Conclusion

Agricultural producers use pesticides on millions of crop acres, primarily to prevent or manage pest infestations. Their use generates economic benefits for farmers and consumers. However, public concerns since the 1960s about the adverse human health and environmental effects of pesticides have led to tighter regulation of these products. This report examines trends in pesticide use from 1960 to 2008. The report employs a new database compiled from USDA pesticide use surveys supplemented by proprietary data, focusing on 21 selected field and specialty crops.

Pesticide use has changed considerably over the past five decades. Rapid growth characterized the 20 years ending in 1981. The total quantity of pesticides applied to the 21 crops analyzed grew from 196 million pounds of pesticide active ingredients in 1960 to 632 million pounds in 1981. Changes in the active ingredients applied, and small annual fluctuations in the annual pesticide use amid a slight downward trend occurred between 1982 and 2008. All these changes were primarily driven by economic factors that determined crop and input prices and were influenced by pest pressures, crop acreage, agricultural practices, and innovations in pest management systems and practices such as IPM, and regulations.

USDA, the Land Grant Universities, and EPA have collaborated in promoting IPM, including support for four regional IPM centers, to develop new methods to manage pests and educate growers about how to use them. This effort may have reduced reliance on pesticides and influenced use patterns by developing pest management approaches that integrate use of chemical, biological, and cultural controls, including new seed varieties, crop rotation, and other practices, often including pest and weather information and predictive models to use pesticides and other practices more efficiently.

An emerging pest management issue is the development of glyphosate-resistant weed populations associated with the large increase in glyphosate use since the late 1990s,[20] largely due to the widespread adoption of herbicide-tolerant corn, cotton, and soybeans. Glyphosate (which replaced other herbicides more toxic to mammals) accounted for about 50 percent of total herbicide quantity in 2008.[21] Resistance to it has induced growers to use other herbicides more toxic to mammals in conjunction with glyphosate in resistance management strategies. This has increased the quantity of herbicides applied and the total cost of weed control. As a result, Federal, State, and private-sector research and extension analysts are working to increase awareness of the problem and to develop resistance management strategies that will maintain the economic effectiveness and environmental benefits associated with the use of herbicides such as glyphosate.

Another important issue is the development of Bt resistance in western corn rootworm, cotton bollworm, and fall armyworm populations leading to reduced efficacy of Bt corn and Bt cotton recently documented in some U.S. crop fields.

Finally, a broad topic with policy implications for pesticide use is the arrival of important invasive species evidenced by the spread since 2000 of the soybean aphid, originally from East Asia. Another case is the arrival of soybean rust, detected in 2004, which increased the application of strobilurin and triazole fungicides and total soybean fungicide use after 2004. As IPM programs for managing

---

[20]Any herbicide widely used alone over time may result in resistant populations.

[21]Glyphosate has low toxicity to mammals, birds, and fish (Malik et al., 1989; NRC, 2010).

insect pests are built upon the existing pest complexes, the introduction of a significant invasive pest can result in heavy pesticide use in the early years of introduction until the biology and ecology of the pest become known.

References

Ackerman, Frank. 2007. "The Economics of Atrazine," *International Journal Occupational Environmental Health* 13(4): 444-449.

Andrilenas, P.A. 1974. *Farmers' Use of Pesticides in 1971: Quantities.* AER-252. U.S. Department of Agriculture (USDA), Economic Research Service (ERS). July.

Andrilenas, P.A. 1975. *Farmers' Use of Pesticides in 1971 -- Extent of Crop Use.* AER-268. U.S. Dept of Agriculture, Economic Research Service.

Boerboom, C., and M. Owen. 2006. "Facts About Glyphosate-Resistant Weeds," Purdue University Extension, GWC-1, Dec.

Campbell, H.E. 1976. "Estimating the Marginal Productivity of Agricultural Pesticides: The Case of Tree Fruit Farms in the Okanagan Valley," *Canadian Journal of Agricultural Economics* 24:23-30.

Carrasco-Tauber, C., and L.J. Moffit. 1992. "Damage Control Econometrics: Functional Specification and Pesticide Productivity," *American Journal of Agricultural Economics* 74(1): 158-162

Carpenter, J., and L. Gianessi. 1999. "Herbicide Tolerant Soybeans: Why Growers are Adopting Roundup Ready Varieties," *AgBioForum* 2:65-72.

Chambers, R.G., and E. Lichtenberg. 1994. "Simple Econometrics of Pesticide Productivity," *American Journal of Agricultural Economics* 76: 407-17.

Claassen, R., F. Carriazo, J. Cooper, D. Hellerstein, and K. Ueda. 2011. *Grassland to Cropland Conversion in the Northern Plains: The Role of Crop Insurance and Disaster Programs.* ERR-120. U.S. Dept of Agriculture, Economic Research Service.

Cohrssen, J.J., and V. T. Covello. 1989. *Risk Analysis: A Guide to Principles and Methods for Analyzing Health and Environmental Risks.* Nat. Tech. Inform. Serv., U.S. Dept. of Commerce.

Council of Environmental Quality. 1993. *The Twenty-fourth Annual Report of the Council on Environmental Quality*, The White House. http://clinton4.nara.gov/CEQ/reports/1993/toc.html.

Criswell, J.T, K. Shelton, and C. Luper. 2013. "Toxicity of Pesticides," Pesticide Applicator Certification Series, Oklahoma Cooperative Extension Service, EPP-7457.

Crop Protection Handbook, Volume 92. 2006. Meister Media Worldwide. Willoughby, OH.

Delvo, H., M. Hanthorn, P. Andrilenas, D. Piper, and T. Lutton. 1983. "Pesticides," in *Inputs Outlook and Situation Report.* U.S. Department of Agriculture (USDA), Economic Research Service (ERS). IOS-2, Oct.

Eichers, T.R., P.A. Andrilenas, R. Jenkins, and A. Fox. 1968. *Quantities of Pesticides Used by Farmers in 1964.* AER-131, U.S. Department of Agriculture, Economic Research Service. Jan.

Eichers,T.R., P.A. Andrilenas, H. Blake, R. Jenkins, and A. Fox. 1970. *Quantities of Pesticides Used by Farmers in 1966*. 1970. AER-179. U.S. Department of Agriculture (USDA), Economic Research Service (ERS). April.

Eichers, T.R., P.A. Andrilenas, and T.W. Anderson. 1978. *Farmers' Use of Pesticides in 1976*. AER-418. U.S. Department of Agriculture (USDA), Economics, Statistics, and Cooperative Service, Dec.

Extoxnet. 1993. "Pesticide Information Profile: Parathion." Oregon State University. http://extoxnet. orst.edu/pips/parathio.htm.

FAO (Food and Agriculture Organization of the United Nations). 1975. Pest Control Problems (preharvest) Causing Major Losses to World Food Supplies. AGP, Pest/PH75/B31. Rome, FAO.

Federal Register. 1997. "Raw and Processed Food Schedule for Pesticide Tolerance Reassessment." Vol. 62, no. 149, Aug. 4.

Fernandez-Cornejo, J. 2004. *The Seed Industry in U.S. Agriculture: An Exploration of Data and Information on Crop Seed Markets, Regulation, Industry Structure, and Research and Development*. AER-786. U.S. Department of Agriculture, Economic Research Service, Jan.

Fernandez-Cornejo, J., and M. Caswell. 2006. T*he First Decade of Genetically Engineered Crops in the United States*. EIB-11, U.S. Department of Agriculture, Economic Research Service, April. http://www.ers.usda.gov/Publications/eib11/

Fernandez-Cornejo, J., and S. Jans. 1999. *Pest Management in U.S. Agriculture*. AH-717, U.S. Department of Agriculture, Economic Research Service, Aug.

Fernandez-Cornejo, J., and S. Wechsler. 2012. "Revisiting the Impact of Bt Corn Adoption by U.S. Farmers," *Agricultural and Resource Economics Review* 41(3): 377-390.

Fernandez-Cornejo, J., and W. McBride. 2002. *Adoption of Bioengineered Crops*. AER-810 U.S. Department of Agriculture, Economic Research Service.

Fernandez-Cornejo J., and Y. Pho. 2002. "Induced Innovation and the Economics of Herbicide Use," *Economics of Pesticides, Sustainable Food Production, and Organic Food Markets*. D.H. Hall and L.J. Moffitt (eds.). Elsevier Science, Ltd, Oxford, UK.

Fernandez-Cornejo, J., C. Alexander, and R. Goodhue. 2002. "Dynamic Diffusion with Disadoption: The Case of Crop Biotechnology in the U.S.A.," *Agricultural and Resource Economics Review* 31(1): 112-126.

Fernandez-Cornejo, J., S. Jans, and M. Smith. 1998. "Issues in the Economics of Pesticide Use in Agriculture: A Review of the Empirical Evidence, "*Review of Agricultural Economics*: 20(2): 462-488.

Fernandez-Cornejo, J., and S. Jans. 1995. "Quality-Adjusted Price and Quantity Indices for Pesticides," *American Journal of Agricultural Economics* 77: 645-59.

Fernandez-Cornejo, J., C. Hallahan, R. Nehring, S. Wechsler, and A. Grube. 2012. "Conservation Tillage, Herbicide Use, and Genetically Engineered Corps in the U.S.A.: The Case of Soybeans," *AgBioForum* 15(3): 231-241.

Foreman, L. 2001. *Characteristics and Production Costs of U.S. Corn Farms.* SB-974. U.S. Department of Agriculture, Economic Research Service, August.

Fox, A., T.R. Eichers, P. Andrilenas, R. Jenkins, and H. Blake. 1968. *Extent of Farm Pesticide Use on Crops in 1966.* AER-147. U.S. Department of Agriculture, Economic Research Service.

Fuglie, K. 1999. "Conservation Tillage and Pesticide Use in the Cornbelt." *Journal of Agricultural and Applied Economics,* 31(1):133-147.

Gardner, J., R. Nehring, and C. Nelson. 2009. " Genetically Modified Crops and Household Labor Savings in US Crop Production," *AgBioForum* 12: 303-312.

Gianessi, Leonard. 2009. "The Benefits of Insecticide Use: Soybeans." CropLife Foundation, Crop Protection Institute, Washington DC.

Griliches, Z. 1957. "Hybrid Corn: An Exploration in the Economics of Technological Change," *Econometrica* 25: 501-522.

Hartzler, B. 1997. *"Chemical Alternatives to Atrazine in Corn Weed Management Programs."* Agronomy Department, Iowa State University Extension.

Hartzler, B., and M.D.K. Owen. 2009. *2010 Herbicide Guide for Iowa Corn and Soybean Production: The Cost of Convenience: Impact of Weeds on Crop Yields.* Agronomy Department, Iowa State University Extension, Nov.

Horn, D.J. 1988. *Ecological Approach to Pest Management.* New York: The Guilford Press.

Horowitz, J., R. Ebel, and K. Ueda. 2010. *No-Till Farming is a Growing Practice.* U.S. Department of Agriculture, Economic Research Service. EIB-70, Nov.

Kellogg, R., R. Nehring, A. Grube, D.W. Goss, and S. Plotkin. 2002. "Environmental Indicators of Pesticide Leaching and Runoff from Farm Fields," *Agricultural Productivity: Measurement and Sources of Growth.* V. Eldon Ball and G. Norton (eds.). Boston: Kluwer Academic Publishers.

Lansink, A.O., and A. Carpentier. 2001. "Damage Control Productivity: An Input Damage Abatement Approach," *Journal of Agricultural Economics* 52(3): 11-22.

Lichtenberg, E., and D. Zilberman. 1986. "The Econometrics of Damage Control: Why Specification Matters," *American Journal of Agricultural Economics* 68: 407-17.

Lin, B., M. Padgitt, L. Bull, H. Delvo, D. Shank, and H. Taylor. 1995. *Pesticide and Fertilizer Use and Trends in U.S. Agriculture.* AER-717. U.S. Department of Agriculture, Economic Research Service. May.

Livingston, M., and C. Osteen. 2012. "Pest Management," *Agricultural Resources and Environmental Indicators, 2012 Edition.* EIB-98. C. Osteen, J. Gottlieb, and U. Vasavada (eds.), Aug.

Lorenz, E. 2009. "Potential Health Effects of Pesticides." Pesticide Safety Fact Sheets, Pennsylvania State University, College of Agricultural Sciences, Agricultural Research and Cooperative Extension, Pesticide Education Program.

Loux, M.M., D. Doohan, A.F. Dobbels, B. Reeb, W.G. Johnson, and T.R. Legleiter. 2013. *2013 Weed Control Guide for Ohio and Indiana*. Ohio State University Extension and Purdue Extension.

Malik, J.M., G.F. Barry, and G.M. Kishore. 1989. "The Herbicide Glyphosate," *BioFactors*.

Morgan, E.R. 2002. *Crop Profile for Cotton in Mississippi*. Mississippi State University, Mississippi State. http://www.IPMCenters.org/CropProfiles/docs/MScotton.html

Munkvold, G., and D. Gorman. 2006. "Foliar Fungicide Use in Corn," *Crop Insights* 16(6): 1-6.

National Research Council (NRC), Environmental Studies Board. 1975. *Pest Control: An Assessment of Present and Alternative Technologies* (Volume 1). Washington, DC: The National Academies Press.

National Research Council (NRC). 1987. *Regulating Pesticides in Food*. Washington, DC: The National Academies Press.

National Research Council (NRC). 1989. *Alternative Agriculture*. Board on Agriculture. Washington, D.C. The National Academies Press.

National Research Council (NRC). 2002. *Environmental Effects of Transgenic Plants: The Scope and Adequacy of Regulation*. Board on Agriculture and Natural Resources, Committee on Environmental Impacts Associated with Commercialization of Transgenic Plants. Washington, DC: The National Academies Press.

National Research Council (NRC). 2010. *The Impact of Genetically Engineered Crops on Farm Sustainability in the United States*. Board on Agriculture and Natural Resources. Washington DC: The National Academies Press. April.

Nesheim, O.N., F.M. Fishel, and M. Mossler. 2005. "Toxicity of Pesticides." University of Florida, IFAS Extension, Document PI-13.

Oerke, E.-C. 2006. "Crop Losses to Pests." *The Journal of Agricultural Science*, (144) 31-43. doi:10.1017/S0021859605005708.

Osteen, C. 1987. "Energy in Plant Nutrition and Pest Control," *Energy in World Agriculture*. Z.R. Helsel (ed.). Amsterdam: Elsevier.

Osteen, C., and J. Fernandez-Cornejo. 2013. "Economic and Policy Issues of U.S. Agricultural Pesticide Use Trends," *Pest Management Science* 69 (9): 1001-1025, September.

Osteen, C., L. Joe Moffitt, and A.W. Johnson. 1988. "Risk Efficient Action Thresholds for Nematode Management," *Journal of Production Agriculture* 1 (4): 332-38.

Osteen, C., and P. Szmedra. 1989. *Agricultural Pesticide Use Trends and Policy Issues*. AER-622. U.S. Department of Agriculture, Economic Research Service, Sept.

Owen, M.D.K. 2009. *2010 Herbicide Guide for Iowa Corn and Soybean Production: New Options for Weed Management in 2010.* Agronomy Department, Iowa State University Extension, Nov.

Owen, M.D.K (Agronomy Department, Iowa State University). 2010. "Herbicide-Resistant Weeds in Genetically Engineered Crops." Statement before the Subcommittee on Domestic Policy, Committee on Oversight and Government Reform, U.S. House of Representatives, Washington, DC, July 28.

Padgitt, M., D. Newton, R. Penn, and C. Sandretto. 2000. *Production Practices for Major Crops in U.S. Agriculture, 1990-97.* SB-969, U.S. Department of Agriculture, Economic Research Service Aug.

Ragsdale, D.W., D.A. Landis, J. Brodeur, G.E. Heimpel and N. Desneux. 2011. "Ecology and Management of the Soybean Aphid in North America," *Annual Review of Entomology* 56: 375-399.

Rogers, E. 1995. *Diffusion of Innovations*, 4th Edition. Free Press, New York.

Sexton, S.E., Z. Lei, and D. Zilberman. 2007. "The Economics of Pesticides and Pest Control," *International Journal of Environmental and Resource Economics* 1: 271-326.

Skopec, M.P. 1998. "Pesticide Trends in Iowa's Surface and Groundwater, 1980-1995," Iowa DNR, Geological Survey, Des Moines, IA.

Strickler, P.E., and W.C. Hinson. 1962. *Extent of Spraying and Dusting on Farms, 1958 – with Comparisons*, SB-314, U.S. Department of Agriculture, Economic Research Service, May.

Szmedra, P.I. 1991. "Pesticide Use in Agriculture," *Handbook of Pest Management in Agriculture*, Volume I. D. Pimentel (ed.), Boca Raton, FL: CRC Press.

Teague, M.L., and B. Wade Brorsen. 1995. "Pesticide Productivity: What Are the trends?" *Journal of Agricultural and Applied Economics* 27: 276-282.

Texas AgriLife Extension Service. 2009. *Crop Profile for Cotton in Texas*, College Station, San Angelo and Lubbock, Texas. http://www.ipmcenters.org/CropProfiles/docs/TXcotton.pdf

Thompson, C. 2011. Personal communication, Agronomy Department, Kansas State University.

U.S. Department of Agriculture (USDA), Economic Research Service (ERS). 2010. *Agricultural Productivity in the United States: Data Documentation and Methods.* http://www.ers.usda.gov/Data/agproductivity/methods.htm

U.S. Department of Agriculture (USDA), Economic Research Service (ERS). 1984. *Inputs Outlook and Situation Report.* IOS-6, Nov.

U.S. Department of Agriculture (USDA), Economic Research Service (ERS). 1986. *Agricultural Resources: Inputs Situation and Outlook Report.* AR-1, Feb.

U.S. Department of Agriculture (USDA), Economic Research Service (ERS). 1987.*Agricultural Resources: Inputs Situation and Outlook Report.* AR-5, Jan.

U.S. Department of Agriculture (USDA), Economic Research Service (ERS). 1988. *Agricultural Resources: Inputs Situation and Outlook Report.* AR-9, Jan.

U.S. Department of Agriculture (USDA), Economic Research Service (ERS). 1989a. *Agricultural Resources: Inputs Situation and Outlook Report.* AR-13, Feb.

U.S. Department of Agriculture (USDA), Economic Research Service (ERS). 1989b. *Agricultural Resources: Inputs Situation and Outlook Report.* AR-15, Aug.

U.S. Department of Agriculture (USDA), Economic Research Service (ERS). 1990a. *Agricultural Resources: Inputs Situation and Outlook Report.* AR-17, Feb.

U.S. Department of Agriculture (USDA), Economic Research Service (ERS). 1990b. *Agricultural Resources: Inputs Situation and Outlook Report.* AR-20, Oct.

U.S. Department of Agriculture (USDA), Economic Research Service (ERS). Agricultural Productivity in the United States: Data Documentation and Methods. 2010. http://www.ers.usda.gov/Data/agproductivity/methods.htm

U.S. Department of Agriculture (USDA) Economic Research Service (ERS). 2010a. "Potatoes: Background." *Briefing Room: Potatoes.* http://www.ers.usda.gov/Briefing/Potatoes/Background.htm

U.S. Department of Agriculture (USDA), Economic Research Service (ERS). 2010b. "Pesticide Expenditures by State: 1949-2009." *Farm Income: Data Files.* http://www.ers.usda.gov/data/FarmIncome/FinfidmuXls.htm

U.S. Department of Agriculture (USDA), National Agricultural Statistics Service (NASS). 2012. "Agricultural Chemical Use Database." http://www.pestmanagement.info/nass/act_dsp_usage_multiple.cfm

U.S. Department of Agriculture (USDA), National Agricultural Statistics Service (NASS). 2012. "Agricultural Prices Summary." http://usda.mannlib.cornell.edu/MannUsda/viewDocumentInfo.do;jsessionid=382330D0A5B78E94538728044886B4D3?documentID=1003

U.S. Department of Agriculture (USDA), National Agricultural Statistics Service (NASS). 2013. "Agricultural Chemical Use Program," http://www.nass.usda.gov/Surveys/Guide_to_NASS_Surveys/Chemical_Use/index.asp

U.S. Department of Agriculture (USDA), National Agricultural Statistics Service (NASS). Various years. *Agricultural Statistics.*

U.S. Department of Agriculture (USDA), National Agricultural Statistics Service (NASS). Various years *Agricultural Chemical Usage: Field Crops Summaries.*

U.S. Department of Agriculture (USDA), National Agricultural Statistics Service (NASS). Various years. *Agricultural Chemical Usage: Fruit Summary.*

U.S. Department of Agriculture (USDA), National Agricultural Statistics Service (NASS). Various years. *Agricultural Chemical Usage: Vegetable Summary.*

U.S. Department of Agriculture (USDA), National Agricultural Statistics Service (NASS). 2011. *Quick-Stats,*

U.S. Department of Agriculture (USDA), National Agricultural Statistics Service (NASS). 2013. *Crop Production Historical Track Records.*

U.S. Environmental Protection Agency (EPA). 1993. R.E.D. Facts: Daminozide," EPA-738-93-007, Sept.

U.S. Environmental Protection Agency (EPA). 1994. "R.E.D. Facts: Mevinphos,"EPA-738-F-94-020, Sept.

U.S. Environmental Protection Agency (EPA). 1995. "R.E.D. Facts: Metolachlor," EPA-738-95-007, April.

U.S. Environmental Protection Agency (EPA). 1998. "R.E.D. Facts: Alachlor," EPA-738-98-018, Dec.

U.S. Environmental Protection Agency (EPA). 1999. *Pesticide Industry Sales and Usage: 1996 and 1997 Market Estimates.* Biological and Economic Analysis Division, Office of Pesticide Programs (A.L. Aspelin and A.H. Grube). 73-R-99-001. Nov.

U.S. Environmental Protection Agency (EPA). 2005a. "Maneb Facts," August;

U.S. Environmental Protection Agency (EPA). 2005b. "Mancozeb Facts," Sept; U.S.

U.S. Environmental Protection Agency (EPA). 2005c. "Metiram Facts," August.

U.S. Environmental Protection Agency (EPA). 2006. "Decision Documents for Atrazine: Finalization of Atrazine IRED, and Completion Tolerance Reassessment and Reregistration Process (April 6, 2006) Revised Atrazine IRED (Oct. 31, 2003), Atrazine IRED (Jan. 2003)."

U.S. Environmental Protection Agency (EPA). 2006a. "Report of the Food Quality Protection Act Tolerance Reassessment Progress and Risk Management Decision (TRED) for Acetochlor," March.

U.S. Environmental Protection Agency (EPA). 2006b. "Reregistration Eligibility Decision for Simazine." EPA 738-R-06-008, April.

U.S. Environmental Protection Agency (EPA). 2008. *Pesticides and Food: What the Pesticide Residue Limits Are on Food.* June. http://www.epa.gov/pesticides/food/viewtols.htm

U.S. Environmental Protection Agency (EPA). 2011. *Pesticide Industry Sales and Usage: 2006 and 2007 Market Estimates.* Biological and Economic Analysis Division, Office of Pesticide Programs (A.Grube, D. Donaldson, T. Kiely, and L. Wu). Feb.

U.S. Environmental Protection Agency (EPA). 2013. "Water: Basic Information about Regulated Drinking Water Contaminants and Indicators." http://water.epa.gov/drink/contaminants/basicin-formation/index.cfm.

Wauchope, R.D., T.M. Butler, A.G. Hornsby, P.W. Augustijn-Beckers, and J.P. Burt. 1992. "The SCS/ARS/CES Pesticide Properties Database for Environmental Decision-Making," *Reviews of Environmental Contamination and Toxicology.* Vol 12. New York: Springer-Verlag.

Wescott, P.C. 2007. *Ethanol Expansion in the United States: How Will the Agricultural Sector Adjust?* FDS-07D-0, U.S. Department of Agriculture, Economic Research Service (May).

Wu, J. 1999. "Crop Insurance, Acreage Decisions, and Nonpoint-Source Pollution," *American Journal of Agricultural Economics*, 81: 305-320.

Yudelman, M., Ratta, A. & Nygaard, D. 1998. *Pest Management and Food Production.* Food, Agriculture, and the Environment Discussion Paper 25. International Food Policy Research. Institute. Washington, DC

Appendix 1—Human Health Effects and Pesticide Regulation

EPA's regulatory process influences risks by defining what pesticides and practices are legal. The hazard or risk of using a pesticide depends on toxicity and exposure (Lorenz, 2009). In a general sense, toxicity is the "quality or degree of being poisonous or harmful to plant, animal, or human life" (Cohrssen and Covello, 1989, p. 374). Pesticide-specific requirements—such as crops that can be treated, product formulations, application rates and methods, and protective clothing and equipment—can limit exposure to the pesticide's toxicities.

Toxicity to humans is usually characterized as acute or chronic. The effects of acute toxicity can result from one-time exposure to relatively large amounts and range from skin irritation to death. The effects of chronic toxicity can result from long-term exposure to small amounts, such as routine exposure while mixing, loading, or applying pesticides, or working in fields after application (Criswell et al., 2013; Nesheim, 2005). Additional concerns include pesticide residues in food or drinking water and non-occupational exposure (such as homeowner use), including risks to children. Potential chronic effects include benign or malignant tumors, birth defects, genetic changes, reproductive effects, blood disorders, and endocrine disruption (Criswell, et al., 2013).

By registering new pesticides that meet health, safety, and environmental standards, and cancelling or restricting registrations of pesticides that do not meet them, the regulatory process induces changes in pesticide use, and has reduced overall worker, dietary, drinking water, and non-occupational risks.

The regulatory process addresses risks from acutely toxic pesticides, including some carbamate and organophosphate insecticides, which affect the nervous system and have caused pesticide poisonings (Lorenz, 2009; Criswell et al., 2013). As a result, growers reduced use of such acutely toxic insecticides and rely on less acutely toxic materials.

- For example, among organophosphates, all fruit, nut, and vegetable uses of ethyl parathion were cancelled in 1992, and all uses of mevinphos were cancelled in 1994 (Extoxnet, 1993; EPA, 1994).

- EPA listed all carbamate and organophosphate insecticides in the highest priority for FQPA (Food Quality Protection Act) tolerance reassessment and reregistration in 1997, resulting in many regulatory actions to reduce worker, dietary, and environmental risks (Federal Register, 1997).

- Since 2000, carbamate and organophosphate shares of insecticide use have decreased, while shares of less acutely toxic insecticides, such as pyrethroids (e.g., permethrin, cypermethrin, tralomethrin, and esfenvalerate) and neo-nicotinoids (e.g., imidicloprid, acetamiprid, thiamethoxam, and clothianidin) increased (Osteen and Fernandez-Cornejo, 2013).

- The adoption of BT corn and cotton, which produce a protein toxic to some insect species but with very low acute toxicity to humans, reduced use of synthetic insecticides, whether acutely or chronically toxic.

However, the regulatory process also addresses chronic risks of pesticides that have low acute toxicities. For example:

- The use of the growth regulator, daminozide (trade name Alar), on apples created a major controversy in the 1980s because of dietary carcinogenic risk. A Special Review resulted in voluntary cancellation of all food use registrations in 1989, while reregistration required practices to reduce greenhouse worker risks in 1993 (EPA, 1993).

- EBDC fungicides (maneb, mancozeb, metiram, zineb), used on many fruits and vegetables, raised concerns about carcinogenic, developmental, and thyroid effects during the 1970s (NRC, 1987). From the early 1980s to the mid-2000s, two Special Reviews and reregistration resulted in cancelled food use registrations, application rate reductions and method restrictions, and other requirements to reduce risks to consumers, agricultural workers, and some aquatic and terrestrial species (EPA, 2005 a-c).

Some widely used herbicides, with relatively low acute toxicities, came under regulatory scrutiny in the 1980s and 1990s, among them triazines (atrazine, cyanazine, and simazine) and acetanilides (alachlor and metolachlor). Primary concerns were dietary and drinking water carcinogenic or oncogenic risks, but also worker safety and homeowner risks. The scrutiny resulted in use restrictions, advisories, and drinking water monitoring, which encouraged changes in herbicide use.

- Atrazine, simazine, and alachlor were detected in drinking water and subjected to compliance monitoring under the Safe Drinking Water Act (SDWA) in the 1990s (EPA, 2013).

- Regulatory actions on triazines from the late 1980s to the mid-2000s included cancellation of atrazine use on some crops and of all cyanazine uses; and restrictions in application rates and methods, worker protection requirements, setbacks from wells and waterways, and drinking water monitoring for atrazine and simazine (EPA, 2006 b, c). (EPA reclassified atrazine in 2003 and simazine in 2005 as "not likely to be carcinogenic," but developmental and reproductive effects remained concerns (EPA, 2006 b, c).)

- Regulatory reviews of alachlor and metolachlor in the 1980s and 1990s resulted in advisories and requirements addressing groundwater and worker protection (EPA, 1995 and 1998). Two newer acetanilides, acetochlor and s-metolachlor (a metolachlor isomer applied at lower rates), introduced in the 1990s, largely replaced alachlor and metolachlor use (EPA, 2006 a).

- Increased glyphosate use, associated with adoption of herbicide-tolerant crops since the mid-1990s, reduced herbicide use shares for triazines, acetanilides, and other materials (Osteen and Fernandez-Cornejo, 2013). Glyphosate has low acute toxicity and is viewed as more environmentally benign than many other herbicides, but has been regulated under the Safe Drinking Water Act since 1994 (EPA, 2013).

Appendix 2—Data Sources

This appendix summarizes the pesticide use surveys used to construct the pesticide use database (appendix table 2.1) and the active ingredients included in the study, classified by pesticide type (appendix table 2.2). The pesticide use database focuses on 21 crops: apples, barley, corn, cotton, grapefruit, grapes, lemons, lettuce, peaches, peanuts, pears, pecans, potatoes, oranges, rice, sorghum, soybeans, sugarcane, sweet corn, tomatoes, and wheat, and covers 1960 through 2008.[22]

The data were compiled from pesticide use surveys carried out by USDA's Economic Research Service (ERS) and National Agricultural Statistics Service (NASS) and proprietary data from a market research firm provided to EPA and shared with ERS (proprietary data for short). The ERS surveys were conducted mainly in the 1960s and 1970s and covered selected crops (Strickler and Hinson, 1962; Eichers et al., 1968, 1970; Fox et al, 1968; Andrilenas, 1974, 1975; Eichers et al. (1978); USDA/ERS, 1984, 1986, 1987, 1988, 1989a, 1989b, 1990a, 1990b). The NASS pesticide surveys (USDA/NASS, various years) began in 1990, but not all crops were covered every year. USDA data were used if available for a specific year, crop, State, and active ingredient. When USDA data were not available, proprietary data were used. If neither USDA nor proprietary data were available, estimates were made based on application rates (e.g., pounds of a.i. per acre) of contiguous years (using USDA or proprietary data) and planted acres reported by USDA.[23] ERS did not have access to proprietary data needed to use this method to estimate pesticide quantities beyond 2008.

In some cases, inferences were made. For example, methyl bromide use was inferred for tomatoes in South Carolina (1992-1998); Arkansas, Tennessee, and Georgia (1997); Florida (1997 and 1999); and California (1998, 1999, and 2001). Metam sodium use was imputed for tomatoes in California (1993, 1995, and 1997), potatoes in Idaho (1996, 1998, 2000, 2002, 2004, 2006, and 2007), Oregon (1996 and 1998), and Washington (1996, 1998, 2000, 2002, 2004, 2006, and 2007). Dichloropropene use was inferred for potatoes in Idaho (1996 and 2004), Oregon (1996 and 1998), and Washington (1996, 1998, 2000, 2002, and 2004).

Proprietary data for oranges were dropped in all years that were bordered by NASS data. Linear interpolations of the NASS data were used to infer active ingredient totals in those years. Since no data were available for wheat fungicides for 1990 and 1992, estimates from Lin et al. (1995) were used to account for fungicide use during those years.

The shares of acres treated with pesticides were obtained from the following sources: Strickler and Hinson (1962); Fox et al. (1968); Andrilenas (1975); Eichers et al. (1978); USDA, ERS *Inputs Outlook and Situation* reports, (USDA/ERS, 1984, 1986, 1987, 1988, 1989a, 1989b, 1990a, 1990b); and USDA/NASS, *Agricultural Chemical Usage Summaries* (1991-2008). Since there were no NASS survey data for 2008, shares of acres treated with pesticides were estimated by linear interpolation using 2009 data for wheat and 2010 data for corn, cotton, and potato obtained from NASS Chemical Use Data in Quick Stats 2.0.

---

[22]However, data for shares of acreage treated with pesticides are available for some crops beyond 2008 (corn, cotton, and potatoes for 2010 and wheat for 2009). These data were used to interpolate the 2008 shares when 2008 data were not available.

[23]Most of the data for the early years (1960-1989) are from State level files that were developed to provide information for the ERS productivity accounts. These files contain active ingredient information by year, State, and crop, and are a synthesis of USDA and proprietary data. They also include estimates for selected pesticides.

Main pesticide use surveys used to construct the 1960 - 2008 database

Survey/Report	Source	Crop	Years covered	States or regions covered
Proprietary data	Market research firm	All 21 crops	1987-89, 1994, 1996-2008	All 48 states
Agricultural Chemical Usage: 2007 Field	USDA/NASS	apples	2007	CA, MI, NY, NC, OR, PA, WA
Crops Summary		cotton	2007	AL, AR, CA, GA, LA, MS, MO, NC, SC, TN, TX\
Agricultural Chemical Usage: 2006 Field	USDA/NASS	lettuce	2006	AZ, CA
Crops Summary; Agricultural Chemical		sweetcorn	2006	CA, CO, FL, GA, IL, MI, MN, NJ, NY, NC
Usage: 2006 Vegetable Summaries				OH, OR, PA, TX, WA, WI
		tomatoes	2006	CA, FL, GA, NJ, NC, OH, TN
		rice	2006	AR, CA, LA, MS, MO, TX
		soybeans	2006	AR, IL, IN, IA, KS, KY, LA, MI, MN, MS
				MO, NE, NC, ND, OH, SD, TN, VA, WI
		wheat	2006	CO, ID, IL, KS, MI, MN, MO, MT, NE
				ND, OH, OK, OR, SD, TX, WA
Agricultural Chemical Usage: 2005 Field	USDA/NASS	apples	2005	CA, MI, NY, NC, OR, PA, WA, WI
Crops Summary; Agricultural Chemical		grapefruit	2005	CA, FL, TX
Usage: 2005 Fruit Summaries		grapes	2005	CA,NY,WA
		lemons	2005	CA
		oranges	2005	CA, FL
		peaches	2005	CA, GA, MI, NJ, PA, SC, TX
		pears	2005	CA, OR, WA
		corn	2005	CO, GA, IL, IN, IA, KS, KY, MI, MN, MO
				NE, NY, NC, ND, OH, PA, SD, TX, WI
		cotton	2005	AL, AR, CA, GA, LA, MS, NC, TN, TX
		potatoes	2005	CO, ID, ME, MI, MN, ND, WA, WI
		soybeans	2005	AR, IL, IN, IA, KS, KY, LA, MI, MN, MS
				MO, NE, NC, OH, SD, TN, WI
Agricultural Chemical Usage: 2004 Field	USDA/NASS	lettuce	2004	AZ, FL
Crops Summary; Agricultural Chemical		sweetcorn	2004	CA, FL, GA, IL, MI, MN, NJ, NY, NC

Pesticide Use in U.S. Agriculture: 21 Selected Crops, 1960-2008, EIB-124
Economic Research Service/USDA

Main pesticide use surveys used to construct the 1960 - 2008 database—continued

Survey/Report	Source	Crop	Years covered	States or regions covered
Usage: 2004 Vegetable Summaries				OH, OR, PA, TX, WA, WI
		tomatoes	2004	CA, FL, GA, NJ, OR, TX, WA, WI
		peanuts	2004	AL, FL, GA, NC, TX
		soybeans	2004	AR, IL, IN, IA, KS, MN, MO, NE, ND, OH, SD
		wheat	2004	CO, ID, IL, KS, MI, MN, MO, MT, NE
				ND, OH, OK, OR, SD, TX, WA
Agricultural Chemical Usage: 2003 Field	NASS	apples	2003	CA, MI, NY, NC, OR, PA, WA
Crops Summary; Agricultural Chemical		grapefruit	2003	CA, FL, TX
Usage: 2003 Fruit Summaries		grapes	2003	CA, MI, NY, WA
		lemons	2003	CA
		oranges	2003	CA, FL
		peaches	2003	CA, GA, MI, NJ, PA, SC, TX
		pears	2003	CA, OR, WA
		barley	2003	CA, ID, MN, MT, ND, PA, SD, UT, WA
				WI, WY
		corn	2003	CO, IL, IN, IA, KS, KY, MI, MN, MO,
				NE, NY, NC, ND, OH, PA, SD, TX, WI
		cotton	2003	AL, AZ, AR, GA, LA, MS, MO, NC, SC
				TN, TX
		potatoes	2003	CO, ID, ME, MI, MN, ND, OR, PA
				WA, WI
		sorghum	2003	CO, KS, MO, NE, OK, SD, TX
Agricultural Chemical Usage: 2002 Field	NASS	lettuce	2002	AZ, CA
Crops Summary; Agricultural Chemical		sweetcorn	2002	CA, FL, GA, MI, MN, NJ, NY, NC, OH
Usage: 2002 Vegetable Summaries				OR, WA, WI
		tomatoes	2002	CA, FL, GA, OH, TN
		corn	2002	IL, IN, IA, MN, NE, OH, WI
		soybeans	2002	AR, IL, IN, IA, KS, KY, LA, MD, MI, MN
				MS, MO, NE, NC, ND, OH, SD, TN, VA, WI
		wheat	2002	CO, IL, KS, MN, MO, MT, NE, ND, OH, OK
				TX, WA

Main pesticide use surveys used to construct the 1960 - 2008 database—continued

Survey/Report	Source	Crop	Years covered	States or regions covered
Agricultural Chemical Usage: 2001 Field	NASS	apples	2001	CA, KS, MI, NY, NC, OR, PA, WA
Crops Summary; Agricultural Chemical		grapefruit	2001	CA,FL
Usage: 2001 Fruit Summaries		grapes	2001	CA, KS, MI, NY, WA
		lemons	2001	CA
		oranges	2001	CA, FL
		peaches	2001	CA, GA, MI, NJ, SC
		pears	2001	CA, OR, WA
		corn	2001	CO, GA, IL, IN, IA, KS, KY, MI, MN, MO
				NE, NY, NC, ND, OH, PA, SD, TX, WI
		cotton	2001	AR, CA, GA, LA, MS, NC, TX
		potatoes	2001	ID, ME, MN, ND, OR, WA, WI
		soybeans	2001	AR, IL, IN, IA, MN, MO, NE, OH
Agricultural Chemical Usage: 2000 Field	USDA/ NASS	lettuce	2000	AZ, CA, FL, NJ
Crops Summary; Agricultural Chemical		sweetcorn	2000	CA, FL, GA, IL, KS, MI, MN, NJ, NY,
Usage: 2000 Vegetable Summaries				OR, PA, WA, WI
		tomatoes	2000	CA, GA, KS, MI, NJ, PA
		corn	2000	CO, IL, IN, IA, KS, KY, MI, MN, MO, NE
				NY, NC, ND, OH, PA, SD, TX, WI
		cotton	2000	AL, AZ, AR, GA, LA, MS, MO, NC, TN, TX
		rice	2000	AR, CA, LA, MS, TX
		soybeans	2000	AR, IL, IN, IA, KS, KY, LA, MI, MN, MS, MO
				NE, NC, ND, OH, SD, TN, WI
		wheat	2000	AR, CO, ID, IL, KS, KY, MN, MO, MT, NE
				NC, ND, OH, OK, OR, SD, TX, WA
Agricultural Chemical Usage: 1999 Field	USDA/ NASS	apples	1999	AZ, CA, GA, MI, NJ, NY, NC, OR, PA, SC, WA
Crops Summary; Agricultural Chemical		grapefruit	1999	AZ, CA, FL, TX
Usage: 1999 Fruit Summaries		grapes	1999	CA, IN, MI, NY, OR, PA, WA
		lemons	1999	AZ, CA
		oranges	1999	AZ, CA, FL, TX
		peaches	1999	CA, GA, MI, NJ, NY, NC, PA, SC, TX, WA
		pears	1999	CA, MI, NY, OR, PA, WA

Main pesticide use surveys used to construct the 1960 - 2008 database—continued

Survey/Report	Source	Crop	Years covered	States or regions covered
		corn	1999	CO, IL, IN, IA, KS, KY, MI, MN, MO, NE
				NC, OH, SD, TX, WI
		cotton	1999	AL, AZ, AR, CA, GA, LA, MS, NC, TN, TX
		peanuts	1999	AL, GA, NC, TX
		potatoes	1999	CO, ID, IN, ME, MI, MN, ND, OR, PA
				WA, WI
		soybeans	1999	AR, IL, IN, IA, KS, KY, LA, MI, MN, MS, MO
				NE, NC, OH, PA, SD, TN
		wheat	1999	IN
Agricultural Chemical Usage: 1998 Field	USDA/ NASS	lettuce	1998	AZ, CA, FL, NJ, NY
Crops Summary; Agricultural Chemical		sweetcorn	1998	CA, FL, GA, IL, MI, MN, NJ, NY, NC, OR, TX
Usage: 1998 Vegetable Summaries				WA, WI
		tomatoes	1998	CA, FL, GA, MI, NJ, NY, NC, TX
		barley	1998	Northeast, Northcentral, South, West
		corn	1998	CO, IL, IN, IA, KS, KY, MI, MN, MO, NE, NC
				OH, PA, SD, TX, WI
		cotton	1998	AL, AZ, AR, CA, GA, LA, MS, NC, TN, TX
		potatoes	1998	PA, WI
		sorghum	1998	KS
		soybeans	1998	AR, IL, IN, IA, KS, KY, LA, MI, MN, MS, MO
				NE, NC, OH, SD, TN
		wheat	1998	CA, CO, GA, ID, IL, KS, LA, MN, MS, MO, MT,
				NE, NC, ND, OH, OK, OR, SD, TX, WA
Agricultural Chemical Usage: 1997 Field	USDA/ NASS	apples	1997	CA, GA, MI, NJ, NY, NC, OR, PA, SC, WA
Crops Summary; Agricultural Chemical		grapefruit	1997	CA, FL
Usage: 1997 Fruit Summaries		grapes	1997	CA, MI, NY, OR, PA, WA
		lemons	1997	CA
		oranges	1997	CA, FL
		peaches	1997	CA, GA, MI, NJ, NY, NC, PA, SC, WA
		pears	1997	CA, NY, OR, PA, WA

Pesticide Use in U.S. Agriculture: 21 Selected Crops, 1960-2008, EIB-124
Economic Research Service/USDA

Main pesticide use surveys used to construct the 1960 - 2008 database—continued

Survey/Report	Source	Crop	Years covered	States or regions covered
		Corn	1997	IL, IN, IA, MI, MN, MO, NE, OH, SD, WI
		Cotton	1997	AL, AZ, AR, CA, GA, LA, MS, MO, NC, SC
				TN, TX
		Potatoes	1997	ID, ME, MN, ND, OR, WA, WI
		Soybeans	1997	AR, DE, IL, IN, IA, KS, KY, LA, MI, MN, MS
				MO, NE, NC, OH, PA, SD, TN, WI
		Wheat	1997	CO, ID, IL, KS, MN, MO, MT, NE, ND, OH, OK, OR
				PA, SD, TX, WA
Agricultural Chemical Usage: 1996 Field	USDA/ NASS	lettuce	1996	AZ, CA
Crops Summary; Agricultural Chemical		sweetcorn	1996	CA, FL, GA, IL, KS, MI, MN, NJ, NY,
Usage: 1996 Vegetable Summaries				OR, WA, WI
		tomatoes	1996	CA, FL, GA, NJ, NC, TX
		Corn	1996	IL, IN, IA, KS, KY, MI, MN, MO, NE, NC
				OH, PA, SC, SD, TX, WI
		Cotton	1996	AZ, AR, CA, GA, LA, MS, TN, TX
		Potatoes	1996	ID, ME, MN, ND, WA
		Soybeans	1996	AR, IL, IN, IA, LA, MN, MS, MO, NE, OH
				TN, WI
		Wheat	1996	CO, ID, KS, MN, MT, NE, ND, OK, OR, SD
				TX, WA
Agricultural Chemical Usage: 1995 Field	USDA/ NASS	apples	1995	CA, GA, MI, NJ, NY, OR, PA, SC, WA
Crops Summary; Agricultural Chemical		grapefruit	1995	CA, FL
Usage: 1995 Fruit Summaries		grapes	1995	CA, MI, NY, OR, PA, WA
		lemons	1995	CA
		oranges	1995	CA, FL
		peaches	1995	CA, GA, MI, NJ, NY, NC, PA, SC, WA
		pears	1995	CA, NY, OR, WA
		Corn	1995	DE, GA, IL, IN, IA, KS, KY, MI, MN, MO
				NE, NC, OH, PA, SD, TX, WI
		Cotton	1995	AZ, AR, CA, LA, MS, TX

Pesticide Use in U.S. Agriculture: 21 Selected Crops, 1960-2008, EIB-124
Economic Research Service/USDA

Main pesticide use surveys used to construct the 1960 - 2008 database—continued

Survey/Report	Source	Crop	Years covered	States or regions covered
		Potatoes	1995	CO, ID, ME, MI, MN, NY, ND, OR, PA,WA, WI
		Soybeans	1995	AR, GA, IL, IN, IA, KY, LA, MN, MS, MO
				NE, NC, OH, TN
		Wheat	1995	CO, ID, IL, KS, MN, MO, MT, NE, ND, OK, OR, SD
Agricultural Chemical Usage: 1994 Field	USDA/ NASS	lettuce	1994	AZ, CA, FL, NJ, NY
Crops Summary; Agricultural Chemical		sweetcorn	1994	CA, FL, GA, IL, MI, MN, NJ, NY, NC,
Usage: 1994 Vegetable Summaries				OR, TX, WA, WI
		tomatoes	1994	CA, FL, GA, MI, NJ, NY, NC, TX
		Corn	1994	IL, IN, IA, MI, MN, MO, NE, OH, SD, WI
		Cotton	1994	AZ, AR, CA, LA, MS, TX
		Potatoes	1994	CO, ID, ME, MI, MN, NY, ND, OR, PA, WA, WI
		Soybeans	1994	AR, DE, IL, IN, IA, MN, MO, NE, OH
		Wheat	1994	CO, ID, IL, KS, MN, MO, MT, NE, ND, OH, OK
				OR, SD, TX, WA
Agricultural Chemical Usage: 1993 Field	USDA/ NASS	apples	1993	CA, MI, NJ, NC, OR, PA, SC, WA
Crops Summary; Agricultural Chemical		grapefruit	1993	CA, FL
Usage: 1993 Fruit Summaries		grapes	1993	CA, MI, NY, OR, PA, WA
		lemons	1993	CA
		oranges	1993	CA, FL
		peaches	1993	CA, MI, NY, NC, PA, SC, WA
		pears	1993	CA, NY, OR, WA
		Corn	1993	GA, IL, IN, IA, KS, KY, MI, MN, MO, NE, NC
				OH, PA, SD, TX, WI
		Cotton	1993	AZ, AR, CA, LA, MS, TX
		Potatoes	1993	CO, ID, ME, MI, MN, NY, ND, OR, PA, WA, WI
		Soybeans	1993	AR, GA, IL, IN, IA, KS, KY, LA, MN, MS, MO
				NE, NC, OH, SD, TN
		Wheat	1993	CO, ID, IL, KS, MN, MT, NE, ND, OH, OK
				OR, SD, TX, WA

Main pesticide use surveys used to construct the 1960 - 2008 database—continued

Survey/Report	Source	Crop	Years covered	States or regions covered
Agricultural Chemical Usage: 1992 Field	USDA/ NASS	lettuce	1992	AZ, CA, NJ, MI, NJ, NY, TX
Crops Summary; Agricultural Chemical		sweetcorn	1992	CA, FL, GA, IL, MI, MN, NC, NJ, NY,
Usage: 1992 Vegetable Summaries				OR, TX, WA, WI
		tomatoes	1992	CA, FL, GA, MI, NJ, NY, NC, TX
		Corn	1992	GA, IL, IN, IA, KS, KY, MI, MN, MO, NE, NC
				OH, PA, SC, SD, TX, WI
		Cotton	1992	AZ, AR, CA, LA, MS, TX
		Potatoes	1992	CO, ID, ME, MI, MN, NY, ND, OR, PA, WA, WI
		Rice	1992	AR, LA
		Soybeans	1992	AR, GA, IL, IN, IA, KS, KY, LA, MN, MS
				MO, NE, NC, OH, SD, TN
		Wheat	1992	AR, CO, ID, IL, IN, KS, MO, MN, MT, ND, NE, OH
				OK, OR, SD, TX, WA
Agricultural Chemical Usage: 1991 Field	USDA/ NASS	apples	1991	AZ, GA, MI, NY, NC, OR, PA, SC, VA, WA
Crops Summary; Agricultural Chemical		grapefruit	1991	AZ, FL
Usage: 1991 Fruit Summaries		grapes	1991	AZ, MI, NY, OR, PA, TX, WA
		lemons	1991	AZ
		oranges	1991	AZ, FL
		peaches	1991	GA, MI, NY, NC, PA, SC, TX, VA, WA
		pears	1991	OR, WA
		Corn	1991	GA, IL, IN, IA, KS, KY, MI, MN, MO, NE, NC
				OH, PA, SC, SD, TX, WI
		Cotton	1991	AZ, AR, CA, LA, MS, TX
		Peanuts	1991	GA, TX, NC, VA
		Potatoes	1991	CO, ID, ME, MI, MN, NY, ND, OR, PA, WA, WI
		Rice	1991	AR, LA
		Sorghum	1991	KS, NE, TX
		Soybeans	1991	AR, GA, IL, IN, IA, KS, KY, LA, MN, MS
				MO, NE, NC, OH, SD, TN
		Wheat	1991	AR, CO, ID, IL, IN, KS, MN. MO, MT, ND, NE, OH,
				OH, OR, SD, TX, WA

Main pesticide use surveys used to construct the 1960 - 2008 database—continued

Survey/Report	Source	Crop	Years covered	States or regions covered
Agricultural Chemical Usage: 1990 Field	USDA/ NASS	Lettuce	1990	CA
Crops Summary; Agricultural Chemical		Sweetcorn	1990	CA
Usage: 1990 Vegetable Summaries		Tomatoes	1990	CA
		Corn	1990	CO, IL, IN, IA, KS, KY, MI, MN, MO, NE, NY
				NC, ND, OH, PA, SC, SD, TN, VA, WV, WI
				Reg 1, Reg 2, Reg 3, Reg 4
		Cotton	1990	AZ, AR, CA, LA, MS, TX
		Potatoes	1990	CO, ID, ME, MI, MN, NY, ND, OR, PA, WA, WI
		Rice	1990	AR, LA
		Soybeans	1990	AL, AR, DE, FL, GA, IL, IN, IA, KS, KY, LA
				MD, MI, MN, MS, MO, NE, NJ, NC, ND, OH
				OK, PA, SC, SD, TN, TX, VA, WI
		Wheat	1990	AR, CO, ID, IL, KS, MO, MN, MT, NE, ND, OH, OK
				SD, TX, WA
Proprietary data	Market research firm	Corn	1970-86	Various regions; IL, IN, IA, MN, MO
				NE, OH, WI, SD, MI
		Soybeans	1970-85	Various regions; IL, IN, IA, MN, MO
				NE, OH, WI
		Cotton	1970-85	Various (four) regions
		Wheat	1973-86	Various (three) regions
		Sorghum	1973-86	Various (three) regions
Pesticides, Outlook and Situation (IOS-2)	USDA/ ERS	Corn	1982	MD, NY, PA, KY, NC, TN, VA, AL, FL, GA ID, AZ
(Delvo et al., 1983)				SC, AR, LA, MS, IL, IA, IN, MO, OH, MI, MN
				WI, KS, NE, SD, ND, OK, TX, MT, WA
		Soybeans	1982	MN, IA, MO, IL, IN, OH, KY, AR, LA, MS, MD
				GA, SC, NC, VA, FL, MI, WI, OK, TX
		Cotton	1982	AL, GA, SC, AR, LA, MS, TN, OK, TX, MO, NC, AZ

Main pesticide use surveys used to construct the 1960 - 2008 database—continued

Survey/Report	Source	Crop	Years covered	States or regions covered
		Sorghum	1982	KY, NC, TN, VA, AL, GA, SC, AR, LA, MS, IL
				MO, KS, NE, SD, OK, TX, AZ
		Wheat	1982	MD, KY, NC, TN, VA, AL, GA, SC, AR, LA
				MS, IL, IA, MO, OH, MI, MN, WI, KS, NE, ND
				SD, OK, TX, ID, MT, WA, AZ
		Rice	1982	AR, LA, MS, MO, TX
		Peanuts	1982	NC, VA, AL, GA, SC, TX
		Barley	1982	NC, VA, SC, MI, MN, WI, ND, SD, ID, MT, WA, AZ
		Oats	1982	NY, PA, NC, SC, AR, OH, MI, MN, WI, ND
				SD, OK, TX, ID, MT, WA
Farmers' Use of Pesticides in 1976	ERS	Corn	1976	National; USDA regions NE, LK, CB, NP, AP
(Eichers et al, 1978)				SE, DT, SP, MT, PA
		Soybeans	1976	National; USDA regions NE, LK, CB, NP, AP
				SE, DT, SP, PA
		Cotton	1976	National; USDA regions AP, SE, DT, SP, MT, PA
		Sorghum	1976	National; USDA regions NE, LK, CB, NP, AP
				SE, DT, SP, MT, PA
		Wheat	1976	National; USDA regions NE, LK, CB, NP, AP
				SE, DT, SP, MT, PA
		Rice	1976	National; USDA regions DT, SP, PA
		Peanuts	1976	National; USDA regions AP, SE, DT, SP, MT
		Barley	1976	National; USDA regions NE, LK, CB, NP, AP
				SE, SP, MT, PA
Proprietary data	Market research firm	Oranges	1974, 1975, 1984	Fl, CA & AZ, TX & national
		Lemons	1974,1975, 1984	Fl, CA & AZ, TX & national
		Grapefruit	1974, 1975, 1984	Fl, CA&AZ, TX & national
		Grapes	1974,1975, 1984	Northeast, CA & AZ & national

Main pesticide use surveys used to construct the 1960 - 2008 database—continued

Survey/Report	Source	Crop	Years covered	States or regions covered
		Apples	1975, 1975, 1984	East, North Central, West & national
		Peaches	1974, 1975, 1984	East, West & national
		Pecans	1974,1975, 1984	Southeast, Southwest & national
		Pears	1974, 1975, 1984	Michigan, Northeast, West & National
Proprietary data (speciatlty crops pesticide study)	Market research firm	Peanuts	19771985	Regional and national
		Tomatoes	1977, 1985	Regional and national
		Sweet corn	1977, 1985	Regional and national
		Potatoes	1977. 1985	Regional and national
		Rice	1977, 1985	Regional and national
Proprietary data on herbicide use	Market research firm	Corn	1961-72	National
		Soybeans	1961-72	National
		Cotton	1961-72	National
		Sorghum	1961-72	National
		Wheat	1961-72	National
		Barley	1961-72	National
Proprietary data on insecticide use	Market research firm	Corn	1961-72	National
		Soybeans	1961-72	National
		Cotton	1961-72	National
		Sorghum	1961-72	National
		Wheat	1961-72	National
		Barley	1961-72	National
Farmer's Use of Pesticides in 1971: Quantities (Andrilenas, 1974)	USDA/ ERS	Corn	1971	National; USDA regions NE, LK, CB, NP, AP SE, DT, SP, MT, PA
		Soybeans	1971	National; USDA regions NE, LK, CB, NP, AP SE, DT, SP
		Cotton	1971	National; USDA regions CB, AP, SE, DT, SP MT, PA
		Sorghum	1971	National; USDA regions NE, CB, NP, AP, SE DT, SP, MT, PA
		Wheat	1971	National; USDA regions NE, LK, NP, AP, SE SP, MT, PA
		Rice	1971	National; USDA regions DT, SP, PA

Main pesticide use surveys used to construct the 1960 - 2008 database—continued

Survey/Report	Source	Crop	Years covered	States or regions covered
		Peanuts	1971	National; USDA regions AP, SE, DT, SP, MT
		Potatoes	1971	National; USDA regions NE, LK, NP, SE, SP,
				MT, PA
		Barley	1971	National; USDA regions NE, LK, CB, NP, SE
				SP, MT, PA
		Apples	1971	National
		Other deciduous fruits	1971	National
		Citrus	1971	National
		Otrher fruits and nuts	1971	National
Quantities of Pesticides Used by	USDA/ERS	Corn	1966	National; USDA regions NE, LK, CB, NP, AP
Farmers in 1966 (Eichers et al., 1970)				SE, DT, SP, MT, PA
		Soybeans	1966	National; USDA regions NE, LK, CB, NP, AP
				SE, DT, SP
		Cotton	1966	National; USDA regions CB, AP, SE, DT, SP
				MT, PA
		Sorghum	1966	National; USDA regions NE, CB, NP, AP, SE
				DT, SP, MT, PA
		Wheat	1966	National; USDA regions LK, CB, NP, AP, SE
				SP, MT, PA
		Rice	1966	National; USDA regions DT, SP
		Peanuts	1966	National; USDA regions AP, SE, SP, MT
		Potatoes	1966	National; USDA regions NE, LK, CB, SE, MT, PA
		Barley	1966	National; USDA regions NE, LK, CB, NP, AP
				SE, SP, MT, PA
		Apples	1966	National
		Other deciduous fruits	1966	National
		Citrus	1966	National

Pesticide Use in U.S. Agriculture: 21 Selected Crops, 1960-2008, EIB-124
Economic Research Service/USDA

Main pesticide use surveys used to construct the 1960 - 2008 database—continued

Survey/Report	Source	Crop	Years covered	States or regions covered
		Other fruits and nuts	1966	National
Quantities of Pesticides Used by	USDA/ERS	Corn	1964	National; USDA regions NE, LK, CB, NP, AP
Farmers in 1964 (Eicher et al., 1968)				SE, DT, SP, MT, PA
		Soybeans	1964	National; USDA regions NE, LK, CB, NP
				AP, SE, DT
		Cotton	1964	National; USDA regions CB, AP, SE, DT, SP, PA
		Sorghum	1964	National; USDA regions CB, NP, AP, SE, SP, MT
		Wheat	1964	National; USDA regions LK, CB, NP, AP, SP
				MT, PA
		Rice	1964	National; USDA regions CB, DT, SP
		Peanuts	1964	National; USDA regions AP, SE, SP, MT
		Potatoes	1964	National; USDA regions NE, LK, CB, SE, MT, PA
		Barley	1964	National; USDA regions NE, LK, CB, NP, AP
				SE, DT, SP, MT, PA
		Apples	1964	National
		Other decisuous fruts	1964	National
		Citrus	1964	National
		Other fuits and nuts	1964	National
Extent of Spraying and Dusting on Farms,	USDA/ERS	Corn	1958	National; USDA regions NE, LK, CB, NP, AP
1958 with Comparisons (Strickler and Hinson, 1962)				SE, DT, SP, MT, PA
		Cotton	1958	National; USDA regions CB, AP, SE, DT, SP, PA
		Potatoes	1958	National; USDA regions NE, LK, CB, SE, MT, PA

Note: USDA Regions are as follows: NE=Northeast, LK=Lake States, CB=Corn Belt, NP=Northern Plains, AP=Appalachian, SE=Southeast, DT=Delta States, SP=Southern Plains, MT=Mountain, and PA=Pacific.

Appendix table 2.2
Active ingredient classifications

Herbicides	Insecticides	Fungicides	Other
2,4,5-T	Abamectin	Acibenzolar S Methyl	6 Benzyladenine
2,4-D	Ac303630	Ampelomyces quisqualis	Arsenic Acid
2,4-Db	Acephate	Anilazine	Aviglycine
Acetic Acid	Acequinocyl	Azoxystrobin	Benzyladenine
Acetochlor	Acetamiprid	Bacillus cereus	Butanone
Acifluorfen	Aldicarb	Bacillus pumilis	Butifos
Alachlor	Aldrin	Bacillus subtilis	Calcium Chloride
Allidochlor	Amitraz	Barium Polysulfide	Chlorophacinone
Ametryn	Azadirachtin	Benomyl	Chloropicrin
Aminopyralid	Azinphos Methyl	Bordeaux Mixture	Cppu
Amitrole	Bacillus thuringiensis	Boscalid	Cyanamide
Asulam	Bifenazate	Burkholderia cepacia	Cyclanilide
Atrazine	Bifenthrin	Calcium Polysulfide	Cytokinin
Barban	Buprofezin	Captan	Daminozide
Benefin	Carbaryl	Carboxin	Dazomet
Benfluralin	Carbofuran	Chinomethionat	Decan 1 Ol
Bensulfuron Methyl	Chlorantraniliprole	Chloroneb	Dichloropropene
Bensulide	Chlordane	Chlorothalonil	Dimethipin
Bentazone	Chlordimeform	Coniothyrium minitans	Dimethylarsinic Acid
Bispyribac Sodium	Chlorethoxyfos	Copper compounds	DNOC
Bromacil	Chlorfenapyr	Cyazofamid	Dodecadien 1 Ol
Bromoxynil	Chlorobenzilate	Cymoxanil	Dodecanol
Bromoxynil Heptanoate	Chlorpyrifos	Cyprodinil	Endothall
Bromoxynil Octanoate	Cinnamaldehyde	Dicloran	Ethephon
Butralin	Clofentezine	Difenoconazole	E E 8 10 Dodecadien
Butylate	Clothianidin	Dimethomorph	Gamma Aminobutyric Acid (GABA)
Carfentrazone Ethyl	Cryolite	Dinocap	Gibberellic Acid
Chloramben	Cydia pomonella granulosis virus	Dodine	Hydrogen Cyanamide
Chloridazon	Cyfluthrin	Etridiazole	Hydroxypropanoic Acid
Chlorimuron Ethyl	Cypermethrin	Famoxadone	Indolyl Butyric Acid
Chlorpropham	Cyromazine	Fenamidone	L Glutamic Acid
Chlorsulfuron	DDT	Fenarimol	Maleic Hydrazide
Clethodim	Deltamethrin	Fenbuconazole	Mepiquat Chloride
Clodinafop Propargyl	Demeton	Fenhexamid	Metaldehyde
Clomazone	Diazinon	Ferbam	Metam Potassium
Clopyralid	Dicofol	Fluazinam	Metam Sodium
Cloransulam Methyl	Dicrotophos	Fludioxonil	Methyl Bromide
Cyanazine	Dienochlor	Fluoxastrobin	Methyl Iodide
Cycloate	Diflubenzuron	Flusilazole	Methyl Isothiocyanate
Cyhalofop Butyl	Dimethoate	Flutolanil	Naphthalene
Dalapon	Dinotefuran	Flutriafol	Naphthylacetamide
DCPA	Disulfoton	Folpet	Naphthylacetic Acid
Desmedipham	Emamectin Benzoate	Fosetyl Aluminium	Octadecadien (E Z)
Dicamba	Endosulfan	Xilenol-+meta-cresol (Gallex)	Octadecadien (Z Z)

Pesticide Use in U.S. Agriculture: 21 Selected Crops, 1960-2008, EIB-124
Economic Research Service/USDA

Active ingredient classifications—continued

Herbicides	Insecticides	Fungicides	Other
Dichlobenil	EPN	Gentamicin Sulfate	Paclobutrazol
Dichlorprop	Esfenvalerate	Harpin Protein	Pelargonic Acid
Diclofop Methyl	Ethion	Hydrogen Peroxide	Phosphoric Acid
Diclosulam	Ethoprophos	Iprodione	Pinolene
Difenzoquat	Etoxazole	Kresoxim Methyl	Piperonyl Butoxide
Diflufenzopyr	Fenamiphos	Mancozeb	Prohexadione Calcium
Dimethenamid	Fenbutatin Oxide	Mandipropamid	Sodium Chlorate
Dimethenamid P	Fenoxycarb	Maneb	Sodium Metaborate
Dinitro	Fenpropathrin	Mefenoxam	Strychnine
Dinoseb	Fenpyroximate	Metalaxyl	Sulfcarbamide
Diphenamid	Fenvalerate	Metconazole	Tetradecanol
Diquat Dibromide	Fipronil	Metiram	Tetrathiocarbonate
Diuron	Flonicamid	Myclobutanil	Thidiazuron
DSMA	Flubendiamide	Oxytetracycline	Tetrathiocarbonate
EPTC	Flucythrinate	Potassium Bicarbonate	Thidiazuron
Ethalfluralin	Fluvalinate	Propamocarb Hydrochloride	Tribufos
Ethofumesate	Fluvalinate Tau	Propiconazole	Zinc
Fenoxaprop Ethyl	Fonofos	Prothioconazole	Zinc Phosphide
Fenoxaprop P Ethyl	Formetanate Hydrochloride	Pyraclostrobin	Z 8 Dodecen Acetate
Florasulam	Gamma Cyhalothrin	Pyrimethanil	
Fluazifop Butyl	Garlic Oil	Quinoxyfen	
Flucarbazone Sodium	Heptachlor	Quintozene	
Fluchloralin	Hexythiazox	Streptomycin	
Flufenacet	Hydramethylnon	TCMTB	
Flumetsulam	Imidacloprid	Tebuconazole	
Flumiclorac Penthyl	Indoxacarb	Tetraconazole	
Flumioxazin	Isofenphos	Thiabendazole	
Fluometuron	Kinoprene	Thiophanate Methyl	
Fluroxypyr Meptyl	Lambda Cyhalothrin	Thiram	
Fluthiacet Methyl	Lindane	Triadimefon	
Fomesafen	Malathion	Triadimenol	
Foramsulfuron	Methamidophos	Trifloxystrobin	
Fosamine	Methidathion	Triflumizole	
Glufosinate Ammonium	Methiocarb	Triforine	
Glyphosate	Methomyl	Triphenyltin Hydroxide	
Glyphosate Trimesium	Methoxychlor	Vinclozolin	
Halosulfuron Methyl	Methoxyfenozide	Zineb	
Hexazinone	Methyl Parathion	Ziram	
Imazamethabenz Methyl	Mevinphos	Zoxamide	
Imazamox	Monocrotophos		
Imazapic	Naled		
Imazapyr	Nosema Locustae Canning		
Imazaquin	Novaluron		

Appendix table 2.2
Active ingredient classifications—continued

Herbicides	Insecticides	Fungicides	Other
Imazethapyr	Oxamyl		
Iodosulfuron	Oxydemeton Methyl		
Isoxaben	Oxythioquinox		
Isoxaflutole	Parathion		
Lactofen	Permethrin		
Linuron	Phorate		
MCPA	Phosalone		
MCPB	Phosmet		
Mecoprop	Phosphamidon		
Mesosulfuron Methyl	Phostebupirim		
Mesotrione	Pirimicarb		
Methazole	Polyhedrosis Virus		
Metolachlor	Potassium Salts		
Metribuzin	Profenofos		
Metsulfuron Methyl	Propargite		
Molinate	Propoxur		
Mono Potassium Salt	Pymetrozine		
MSMA	Pyrethrins		
Napropamide	Pyridaben		
Naptalam	Pyriproxyfen		
Nicosulfuron	Rotenone		
Norflurazon	Ryanodine		
Orthosulfamuron	Sabadilla		
Oryzalin	Spinetoram		
Oxyfluorfen	Spinosyn A		
Paraquat	Spirodiclofen		
Pebulate	Spiromesifen		
Pendimethalin	Spirotetramat		
Penoxsulam	Steinernema carpocapsae		
Phenmedipham	Steinernema riobravis		
Phytophthora Spores	Sulprofos		
Picloram	S Methoprene		
Pinoxaden	Tebufenozide		
Primisulfuron Methyl	Tebupirimfos		
Profluralin	Tefluthrin		
Prometon	Temephos		
Prometryn	Terbufos		
Pronamide	Thiacloprid		
Propachlor	Thiamethoxam		
Propanil	Thiodicarb		
Propazine	Toxaphene		
Propoxycarbazone Sodium	Tralomethrin		
Propyzamide	Trimethacarb		
Prosulfuron	Trichlorfon		

Pesticide Use in U.S. Agriculture: 21 Selected Crops, 1960-2008, EIB-124
Economic Research Service/USDA

Appendix table 2.2
Active ingredient classifications—continued

Herbicides	Insecticides	Fungicides	Other
Pyraflufen Ethyl	Zeta Cypermethrin		
Pyrasulfotole			
Pyridate			
Pyrithiobac Sodium			
Quinclorac			
Quizalofop Ethyl			
Rimsulfuron			
Sethoxydim			
Simazine			
Sodium Arsenite			
Sulfentrazone			
Sulfosate			
Sulfosulfuron			
Tebuthiuron			
Tembotrione			
Terbacil			
Terbutryn			
Thiazopyr			
Thifensulfuron Methyl			
Thiobencarb			
Topramezone			
Tralkoxydim			
Triasulfuron			
Tribenuron Methyl			
Triclopyr			
Tridiphane			
Trifloxysulfuron Sodium			
Trifluralin			
Triflusulfuron Methyl			
Triallate			
Urea Sulfate			
Vernolate			

Source: Economic Research Service with USDA and proprietary data.

Appendix 3—The Economics of Pesticide Use

Economics provides a framework for understanding farmers' pesticide use decisions. To maximize profits, farmers should increase the use of pest control inputs until the marginal value of damage reduction (or value of the marginal product-VMP) equals the marginal factor cost (MFC) of the pest control inputs. The VMP is the value of the incremental output (crop) resulting from an additional unit of input (e.g., pesticide), and is equal to the marginal physical product of an input (incremental quantity of output per additional unit of input) multiplied by the price of the output (e.g., crop). If the VMP exceeds the MFC, it will pay farmers to use additional amounts of pest control inputs (Headley, 1968; Campbell, 1976; Lichtenberg and Zilberman, 1986; Teague and Brorsen, 1995).[24] Estimates of the VMP of pesticides per dollar spent on pesticides vary by pesticide type, crop, region, time of the study and methodology used. Studies conducted for the United States and Europe indicate that VMPs per dollar spent on pesticides usually ranged between $1 and $8 but appear to be falling and are usually lower for herbicides (Fernandez-Cornejo et al., 1998; Lansink and Carpentier, 2001). [25]

The VMP depends on potential yield losses and crop prices. The potential yield losses (damages from pest infestations) depend on the extent, intensity, and variability of the pest infestations and the effectiveness of the pest control inputs. The marginal cost of the pest control inputs includes the cost of pesticides and their application. In addition to the private costs, pesticide use may involve external costs such as pest resistance; destruction of beneficial species; heightened threats to worker safety, food safety, and groundwater contamination; damage to nontarget species; air pollution; and property damage (Sexton et al., 2007). Since these costs are usually not borne by the farmers, government policies attempt to control these external costs by imposing restrictions on pesticide use, transferring some external costs to users, or encouraging alternative practices such as Integrated Pest Management (Fernandez-Cornejo et al., 1998).

Under diminishing returns,[26] the VMP declines as more input is used. The downward sloping portion of the VMP is the farmer's demand curve for pesticides, i.e., a relation between price and quantity demanded. The demand depends on many factors, including the price of the crops, the productivity of the pesticides, and other inputs used. Fernandez-Cornejo et al. (1998) observed that, in general, farmers' response to pesticide price changes, as measured by demand elasticity, is small in the short run and moderate in the long run, meaning that pesticide price changes would have little impact on aggregate pesticide use in the short run. Since multiple pesticides may control the same pests, the change in quantity of individual pesticide ingredients used as their own prices change would depend on the availability and cost of alternative pesticides or practices, and demand for individual pesticides could be more responsive to own-price changes than aggregate pesticide use would be.

---

[24]Since by law pesticides must be applied at label rates, the rule really applies as an inequality: Apply if VMP>=MFC, otherwise do not apply. It is a violation of EPA regulations to apply at rates above the label, and pesticide manufacturers do not guarantee efficacy below label rates. However, label rates apply for each application and for each active ingredient. Farmers may apply several active ingredients and several applications in a year. So in the aggregate, e.g., nationally, the equality may apply.

[25]Methodological and modeling issues include the determination of the correct functional form of the underlying production function, and the interaction with other pesticides (Lichtenberg and Zilberman, 1986; Carrasco-Tauber and Moffitt, 1992; Chambers and Lichtenberg, 1994). As Sexton et al. (2007) observe, since there is little consensus on the proper functional form, "economists must work with agronomists and biologists to arrive at the appropriate functional forms that recognize the underlying biological processes."

[26]As pesticide use increases while other inputs are fixed, after some point, the resulting increases in output become smaller.

A demand curve for pesticide x used on crop y for pest z is the summation of individual per-acre grower demands across all acres of crop y and is a relationship between pesticide price, P_x, and quantity demanded, Q_x (see app. fig. 3.1).[27] The relationship assumes that: (1) the individual grower applies a pesticide at a rate based on VMP or one that is scientifically or legally pre-determined, only if net returns are positive; and (2) if two or more pesticides or practices are available, the grower chooses the one with the highest net return. An optimal quantity per acre for pesticide x (q_x) would be determined by $VMP_x = P_x$, where $VMP_x = f(P_y, V_z, d_z, k_x)$, P_y = crop price, V_z = infestation of pest z, d_z = damage per pest, and k_x = pest reduction from pesticide x. (V_z, d_z, and k_x are based on biological and technical relationships, which could be nonlinear, and reflect such factors as weather, soils, and current and previous crop production practices.) The pesticide would be applied only if net returns (NR_x) are greater than zero, that is, value of damage reduction exceeds cost of the pesticide and application: $NR_x = [P_y(V_z*d_z) k_x] - [P_x(q_x) + B_x] > 0$, where B_x is application cost for pesticide x.

Factors that would shift the demand curve rightward (from D_1 to D_2 to D_3) include increases in crop price (P_y) or infestation (V_z), and decreases in application costs (B_x) or the net return of alternative practices (NR_a) (opposite changes would shift the curve leftward). (The net return of alternative practice "a" is a function of its own price, crop price, pest reduction, and application cost, that

Appendix figure 3.1
Supply and demand curves for pesticide(x) when applied to crop(y) for pest(z)

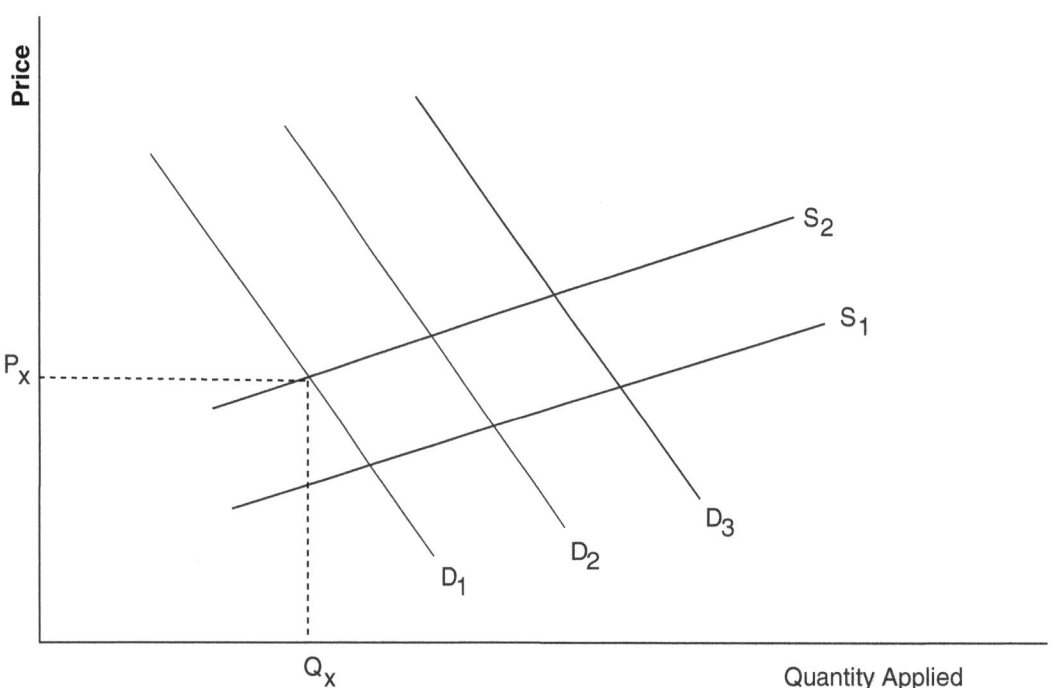

<hr>

[27]Pest management decisions, including pesticide application and timing, are best examined in the context of dynamic, bio-economic models. This simple, static framework does not consider many of the dynamic relationships between crops, pests, and beneficial organisms; the interactions between various practices and natural conditions, or external effects such as the benefits of controlling mobile pests or the costs of adverse environmental or health effects. However, a discussion of dynamic, bio-economic models is beyond the scope of this report.

is $NR_a = f(P_a, P_y, k_a, B_a)$). The shifts are not necessarily proportional to changes in the factor, since the relationships between pest infestations, application rates, and yield can be nonlinear. Changes in labor or fuel prices could affect returns of both pesticide x and the alternatives, by influencing application costs, as could new application technology. Depending on those impacts, increases in labor or fuel prices could shift the demand curve for pesticide x rightward or leftward. Similarly, change in weather could affect yields, infestations, damages, and pesticide effectiveness simultaneously.

The supply curve (a relationship between price and quantity supplied) reflects pesticide marginal costs. It could have a positive slope if competing demands for the material restrict availability as use increases. Changes in pesticide manufacturing cost, regulations that reduce availability, or patent expiration would shift the supply curve. Increased manufacturing cost or regulatory restriction would shift the supply curve upward, increasing P_x and reducing Q_x (shift from S_1 to S_2). Reduced manufacturing costs or patent expiration would shift it downward, with the opposite effects on P_x and Q_x (shift from S_2 to S_1).

Appendix 4—Trends in Pesticide Use on Five Major Crops

This appendix discusses trends and issues for the top five crops—corn, soybeans, cotton, wheat, and potatoes—as measured by total pesticide quantity applied. To facilitate the analysis, we divide the period of study (1960-2008) into two time periods. The first period (1960-1981) is marked by a rapid adoption of pesticides (particularly herbicides) and the second period (1982-2008) is characterized by a slight downward trend with some fluctuations of pesticide use and changes in active ingredients used. For corn, cotton, and soybeans, time period 2 is further divided into 1982-1995 and 1996-2008 to highlight the effect of the adoption of genetically engineered seed on pesticide use (1996 is the year when GE seeds were first commercially introduced).

Average application rates per planted acre (total quantity applied divided by planted acreage) differ by crop, vary over time, and reflect crop-specific rates per treated acre and shares of acreage treated. Among the five crops, potatoes generally had the highest application rates, followed by cotton, corn, soybeans, and wheat (app. figs. 4.1-4.5; app. tables 4.1-4.5). Application rates on potatoes increased from 6.6 pounds per planted acre in 1960 to 44 pounds in 1999 and reached nearly 50 pounds in 2008 (app. fig. 4.4, app. table 4.4). Cotton rates declined sharply from 8-12 pounds per planted acre in the late 1960s through mid-1970s to a low of about 3 pounds by 1980 and fluctuated at 3-5 pounds over the following three decades (app. fig. 4.3, app. table 4.3).

Application rates for all corn pesticides peaked in 1987 at 3.6 pounds per planted acre, declined to 2.1-2.3 pounds per acre in 2000-06, and increased to 2.4 pounds per acre by 2008 (app. fig. 4.1, app. table 4.1). Rates on soybeans peaked at around 2 pounds per planted acre in the early 1980s, declined to 1.1 pounds per acre in the late 1990s and early 2000s, and rose to 1.5 pounds per acre by 2008 (app. fig. 4.2, app. table 4.2). Wheat rates increased from 0.1 pound per planted acre in 1960 to 0.4 pound in 1980, then fluctuated between 0.2 and 0.4 pound from 1980 to 2008 (app. fig. 4.5, app. table 4.5).

Corn Pesticide Trends

Corn pesticide use steadily increased from 29 million pounds in 1960 to 279 million pounds in 1982. It fell below 200 million pounds in 1999-2006, before rising to 204 million pounds in 2008 (app. table 4.1), about 40 percent of the total pesticide pounds applied that year. Corn drives overall pesticide use because it occupies the largest acreage of all U.S. crops. Corn producers apply more pounds of pesticide than any other crop farmers, but use pesticides much less intensively than cotton, fruit, and vegetable producers. For example, potato producers use about 50 pounds per planted acre (app. table 4.4), versus 2.4 pounds per corn acre (app. table 4.1).

1960-1981: Rapid adoption of pesticides and emergence of environmental concerns

One of the first synthetic organic herbicides, 2,4-D, was one of the most widely used on corn in the early 1960s (Eichers et al., 1968). But in 1968, the herbicide atrazine (introduced in 1958) accounted for nearly 60 percent of pesticide use on corn, while 2,4-D accounted for 13 percent (app. fig 4.6). Often combined with alachlor or metolachlor, atrazine was used to control a wide spectrum of broadleaf weeds and grasses (Ackerman, 2007).

Regarding insecticides, corn farmers replaced chlorinated hydrocarbons, such as aldrin and heptachlor (app. figs. 4.6-4.7; see p. 78), which were the subject of environmental and regulatory concerns, with other materials such as carbofuran, phorate, fonofos, and terbufos to control corn root worm and other soil insects (*Crop Protection Handbook*, 2006). Notably, aldrin, introduced in the late 1940s, accounted for 15 percent of total corn pesticide use in 1968, but was no longer used by the late 1970s (app. fig. 4.6).

1982-1995: Fluctuation in pesticide use and changes in active ingredients used

Atrazine was a dominant herbicide during this period. Fueled by large doses of atrazine and the introduction of alachlor and metolachlor, herbicide use reached 3.2 pounds per planted acre during the late 1980s before declining to 2.7 pounds per acre in 1996. Atrazine accounted for more than 25 percent of all corn herbicide use throughout the period (USDA/NASS, *Agricultural Chemical Usage: Field Crop Summaries*, various years). Scrutiny of the potential negative environmental impacts of atrazine and other triazines led EPA, in 1991, to direct States to develop pesticide management plans promoting voluntary reductions in atrazine use, including restrictions on per-acre application rates (Hartzler, 1997; Skopec, 1998).

Though insecticide use rose until the early 1980s, it fell throughout the late 1980s and 1990s. By 1995, terbufos and chlorpyrifos represented more than 60 percent of corn insecticide use (USDA/NASS, *Agricultural Chemical Usage: Field Crop Summaries*, 1996), the bulk of the 0.26 pound of insecticides applied per planted acre (app. table 4.1). Farmers also improved pest control by using crop rotation strategies and by planting traditional hybrid seeds that exhibit better standability.

1996-2008: Adoption of GE corn reduces insecticide use and boosts glyphosate use

The adoption of GE (genetically engineered) corn in the mid-1990s influenced both herbicide and insecticide use. Herbicide use dropped from 2.67 pounds per planted corn acre in 1996 to 1.92 pounds in 2002, hovered below 2.1 pounds between 2003 and 2006, and increased to 2.3 pounds

Appendix figure 4.1
Pounds of pesticide active ingredient (a.i.) per planted acre, corn, 1960-2008

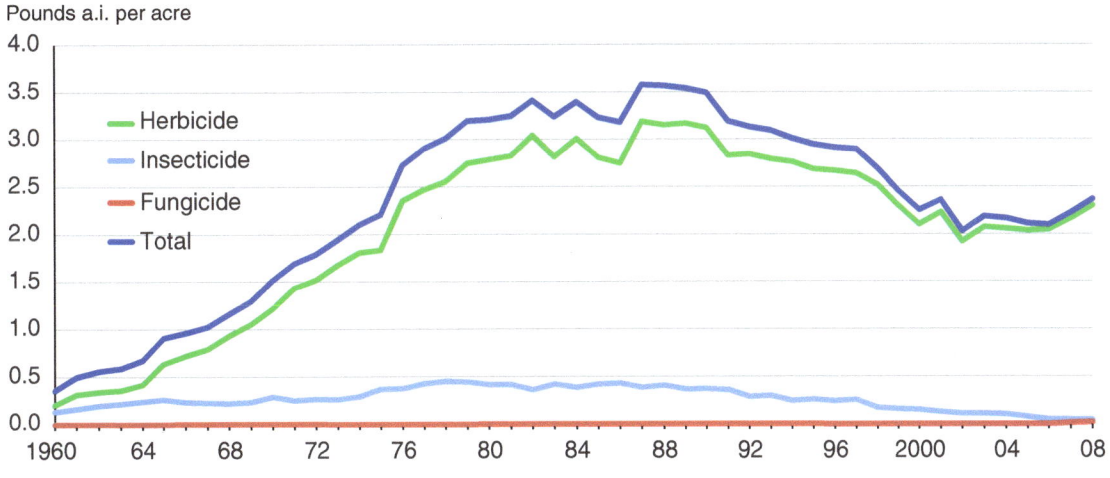

Sources: Economic Research Service with USDA and proprietary data. See Appendix 2.

Pesticide use in corn, 1960-2008

			Pounds active ingredient (a.i.) per planted acre and total pounds a.i. of pesticide applied, by type						
	Millions of planted acres	Herbicide	Insecticide	Fungicide	Total	Herbicide	Insecticide	Fungicide	Total
		Millions of pounds a.i. applied				*Pounds a.i. per planted acre*			
1960	81.4	16.67	11.05	*	29.14	0.20	0.14	ŧ	0.36
1961	65.9	20.87	10.77	*	32.89	0.32	0.16	ŧ	0.50
1962	65.0	22.37	12.92	*	36.38	0.34	0.20	ŧ	0.56
1963	68.8	24.85	14.96	*	40.73	0.36	0.22	ŧ	0.59
1964	65.8	27.79	16.01	*	44.55	0.42	0.24	ŧ	0.68
1965	65.2	41.53	17.13	*	59.26	0.64	0.26	ŧ	0.91
1966	66.3	47.91	15.51	*	63.85	0.72	0.23	ŧ	0.96
1967	71.2	56.47	16.25	*	73.14	0.79	0.23	ŧ	1.03
1968	65.1	60.78	14.49	*	75.68	0.93	0.22	ŧ	1.16
1969	64.3	67.86	15.22	*	83.50	1.06	0.24	ŧ	1.30
1970	66.9	81.77	19.37	*	101.53	1.22	0.29	ŧ	1.52
1971	74.2	106.53	18.79	*	125.70	1.44	0.25	ŧ	1.69
1972	67.1	102.15	17.90	*	120.37	1.52	0.27	ŧ	1.79
1973	72.3	121.18	19.05	*	140.46	1.68	0.26	ŧ	1.94
1974	77.9	141.09	22.86	*	164.12	1.81	0.29	ŧ	2.11
1975	78.7	144.51	29.24	*	173.85	1.84	0.37	ŧ	2.21
1976	84.6	199.21	31.85	*	231.07	2.36	0.38	ŧ	2.73
1977	84.3	208.72	36.30	*	245.05	2.48	0.43	ŧ	2.91
1978	81.7	209.09	36.99	*	246.13	2.56	0.45	ŧ	3.01
1979	81.4	223.94	36.17	*	260.18	2.75	0.44	ŧ	3.20
1980	84.0	234.45	35.22	*	269.74	2.79	0.42	ŧ	3.21
1981	84.1	237.84	35.18	*	273.11	2.83	0.42	ŧ	3.25
1982	81.9	249.40	29.95	*	279.45	3.05	0.37	ŧ	3.41
1983	60.2	169.79	25.38	*	195.26	2.82	0.42	ŧ	3.24
1984	80.5	242.30	31.34	*	273.73	3.01	0.39	ŧ	3.40
1985	83.4	234.57	35.21	*	269.87	2.81	0.42	ŧ	3.24
1986	76.6	210.64	33.09	*	243.82	2.75	0.43	ŧ	3.18
1987	66.2	211.10	25.56	*	236.74	3.19	0.39	ŧ	3.58
1988	67.7	213.75	27.70	*	241.52	3.16	0.41	ŧ	3.57
1989	72.3	229.30	26.67	*	256.04	3.17	0.37	ŧ	3.54
1990	74.2	231.56	27.41	*	259.04	3.12	0.37	ŧ	3.49
1991	76.0	215.10	27.48	NA	242.58	2.83	0.36	NA	3.19
1992	79.3	225.66	22.95	*	248.66	2.85	0.29	ŧ	3.14
1993	73.2	204.80	21.88	*	226.83	2.80	0.30	ŧ	3.10
1994	78.9	218.22	19.39	*	237.64	2.77	0.25	ŧ	3.01
1995	71.5	192.13	18.60	*	210.73	2.69	0.26	ŧ	2.95
1996	79.2	211.44	19.30	*	230.91	2.67	0.24	ŧ	2.91
1997	79.5	210.06	20.24	*	230.43	2.64	0.25	ŧ	2.90

Pesticide Use in U.S. Agriculture: 21 Selected Crops, 1960-2008, EIB-124
Economic Research Service/USDA

Appendix table 4.1

Pesticide use in corn, 1960-2008—continued

	Millions of planted acres	Pounds active ingredient (a.i.) per planted acre and total pounds a.i. of pesticide applied, by type							
		Herbicide	Insecticide	Fungicide	Total	Herbicide	Insecticide	Fungicide	Total
		Millions of pounds a.i. applied				Pounds a.i. per planted acre			
1998	80.2	202.06	13.77	NA	215.84	2.52	0.17	NA	2.69
1999	77.4	177.85	12.16	NA	190.03	2.30	0.16	NA	2.46
2000	79.6	167.53	11.99	NA	179.52	2.11	0.15	NA	2.26
2001	75.7	169.22	9.51	NA	178.74	2.24	0.13	NA	2.36
2002	78.9	151.76	8.61	NA	160.37	1.92	0.11	NA	2.03
2003	78.6	163.21	8.65	NA	171.97	2.08	0.11	NA	2.19
2004	80.9	166.84	8.40	NA	175.48	2.06	0.10	NA	2.17
2005	81.8	166.89	6.14	NA	173.03	2.04	0.08	NA	2.12
2006	78.3	160.60	3.97	NA	164.57	2.05	0.05	NA	2.10
2007	93.5	202.74	4.52	1.21	208.47	2.17	0.05	0.01	2.23
2008	86.0	198.02	4.01	1.69	203.73	2.30	0.05	0.02	2.37

*Indicates that less than 1 million pounds were applied. t indicates less than 0.01. NA indicates not available.
The total columns include other pesticide types, such as defoliants and desiccants.

Source: Economic Research Service with USDA and proprietary data. See Appendix 2.

by 2008 (app. fig. 4.1). The decline until 2002 was driven mainly by the introduction of low-dosage herbicides, such as nicosulfuron.[28]

Though virtually all soybean acres were herbicide-tolerant (HT) by 2000, less than 10 percent of corn acres were. However, use of HT corn seed increased to one-third of corn acres in 2006 and two-thirds in 2008. Consequently, glyphosate accounted for 33 percent of corn pesticide use in 2008, exceeding atrazine use at 29 percent (app. fig. 4.7; see p. 78). In part, the increase in per-acre herbicide use from 2006 to 2008 occurred because the per-acre rates of glyphosate exceeded those of some of the herbicides that it replaced. Also, weed resistance to glyphosate emerged as its use increased, which may have boosted total herbicide use in some States as atrazine was applied along with glyphosate (Hartzler and Owen, 2009).

As a result of increased adoption of Bt corn by U.S. farmers for insect control, insecticide use declined steadily from 0.24 pound per planted acre in 1996 to 0.15 pound by 2000 and 0.05 pound in 2008. Insecticides accounted for only 2 percent of total pesticide pounds applied to corn in 2008 (app. fig. 4.1 and app. table 4.1). However, these estimates do not account for the increased use of insecticide seed treatments with such materials as neo-nicotinoids (for example, imidicloprid, clothianidin, or thiamethoxam), applied at very low rates, because seed treatments are not included in the USDA or proprietary data.

Use of foliar fungicides (applied to the corn plant rather than to soil, seed, or root to control pathogens) was rare prior to 2006 and amounted to less than 1 percent of total corn pesticide use in 2008

---

[28]Despite restrictions, atrazine continued to dominate herbicide use for much of this period (Ackerman, 2007). Aceto-chlor and s-metolachlor (a metolachlor isomer applied at lower rates) were introduced in the 1990s, reducing the market shares of alachlor and the older form of metolachlor. Acetochlor mainly replaced alachlor as a result of a regulatory decision by the registrant, since alachlor was being detected in ground water.

(app. table 4.1) (Owen, 2009). Major diseases controlled by foliar fungicides include gray leaf spot, rusts, and northern leaf blight (Munkvold and Gorman, 2006).

Soybean Pesticide Trends

Pesticide use on soybeans increased from less than 3 million pounds a.i. in 1960 to 146 million pounds in 1982. Usage declined to less than 100 million pounds from 1985 to 2005 and increased to 112 million pounds in 2008 (app. table 4.2). Soybean producers accounted for 22 percent of total pesticide pounds applied in 2008, slightly more than half of the pounds applied in corn production (app. table 4.1). In most years, insecticides and foliar fungicides are applied to 10 percent or less of soybean acreage, limiting per-acre use to less than 0.1 pound.

1960-1981: Rapid increase in pesticide use

Herbicide use on soybeans increased from less than 0.1 pound per acre in 1960 to 2.1 pounds by 1981 (app. fig. 4.2 and app. table 4.2), replacing tillage and cultivation practices as primary weed controls. Chloramben, which is used to control grasses and broadleaf weeds, was the first product to dominate the soybean herbicide market, accounting for over 30 percent of all pesticide use on soybeans in 1968 (app. fig. 4.8). Over time, chloramben use declined, while use of other materials—such as alachlor, metribuzin, and trifluralin—increased (Andrilenas, 1974; Eichers et al., 1970 and 1978).

1982-1995: Fluctuation in pesticide use and changes in active ingredients used

Herbicide per planted soybean acre ranged from 1 to 2 pounds during 1982-95 (app. fig. 4.2 and app. table 4.2). In 1995, pendimethalin, trifluralin, and glyphosate were the most used herbicides in terms of pounds applied (USDA/NASS, *Agricultural Chemical Usage: Field Crop Summaries*, 1996).

Appendix figure 4.2
Pounds of pesticide active ingredient (a.i.) per planted acre, soybeans, 1960-2008

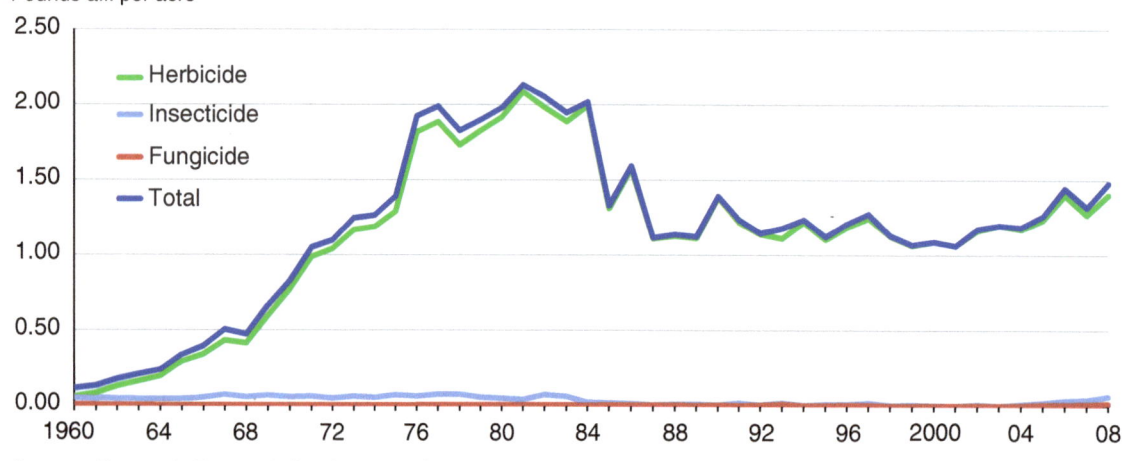

Pounds a.i. per acre

Sources: Economic Research Service with USDA and proprietary data. See Appendix 2.

Pesticide use in soybeans, 1960-2008

	Millions of planted acres	Herbicide	Insecticide	Fungicide	Total	Herbicide	Insecticide	Fungicide	Total
		Millions of pounds a.i. applied				Pounds a.i. per planted acre			
1960	24.4	1.41	1.22	*	2.74	0.06	0.05	ŧ	0.11
1961	27.8	2.27	1.25	*	3.61	0.08	0.04	ŧ	0.13
1962	28.4	3.60	1.29	*	4.97	0.13	0.05	ŧ	0.17
1963	29.5	4.77	1.34	*	6.16	0.16	0.05	ŧ	0.21
1964	31.7	6.13	1.39	*	7.57	0.19	0.04	ŧ	0.24
1965	35.2	10.32	1.52	*	11.88	0.29	0.04	ŧ	0.34
1966	37.3	12.71	2.03	*	14.79	0.34	0.05	ŧ	0.40
1967	40.8	17.70	2.90	*	20.65	0.43	0.07	ŧ	0.51
1968	42.3	17.56	2.41	*	20.03	0.42	0.06	ŧ	0.47
1969	42.5	25.36	2.82	*	28.25	0.60	0.07	ŧ	0.66
1970	43.1	33.20	2.41	*	35.69	0.77	0.06	ŧ	0.83
1971	43.5	42.95	2.63	*	45.66	0.99	0.06	ŧ	1.05
1972	46.9	48.78	2.26	*	51.54	1.04	0.05	ŧ	1.10
1973	56.5	66.06	3.49	*	70.48	1.17	0.06	ŧ	1.25
1974	52.5	62.30	2.70	*	66.36	1.19	0.05	ŧ	1.26
1975	54.6	70.40	3.73	*	75.91	1.29	0.07	ŧ	1.39
1976	50.3	91.50	3.01	*	96.72	1.82	0.06	ŧ	1.92
1977	59.0	111.11	4.27	*	117.24	1.88	0.07	ŧ	1.99
1978	64.7	112.15	4.70	*	118.36	1.73	0.07	ŧ	1.83
1979	71.4	130.53	3.80	*	135.48	1.83	0.05	ŧ	1.90
1980	69.9	134.20	3.39	*	138.39	1.92	0.05	ŧ	1.98
1981	67.5	140.96	2.61	*	144.02	2.09	0.04	ŧ	2.13
1982	70.9	140.51	4.92	*	145.53	1.98	0.07	ŧ	2.05
1983	63.8	120.35	3.83	*	124.27	1.89	0.06	ŧ	1.95
1984	67.8	135.11	1.49	*	136.69	1.99	0.02	ŧ	2.02
1985	63.1	82.77	1.12	*	83.99	1.31	0.02	ŧ	1.33
1986	60.4	95.21	0.86	*	96.16	1.58	0.01	ŧ	1.59
1987	58.2	64.66	0.23	*	64.97	1.11	0.00	ŧ	1.12
1988	58.8	66.37	0.56	*	67.02	1.13	0.01	ŧ	1.14
1989	60.8	67.76	0.50	*	68.34	1.11	0.01	ŧ	1.12
1990	57.8	80.09	0.31	*	80.49	1.39	0.01	ŧ	1.39
1991	59.2	72.19	1.04	*	73.27	1.22	0.02	ŧ	1.24
1992	59.2	67.51	0.18	*	67.82	1.14	0.00	ŧ	1.15
1993	60.1	66.82	1.15	*	70.79	1.11	0.02	ŧ	1.18
1994	61.6	75.24	0.13	*	76.14	1.22	0.00	ŧ	1.24
1995	62.5	68.99	0.46	*	70.23	1.10	0.01	ŧ	1.12
1996	64.2	76.16	0.48	*	77.62	1.19	0.01	ŧ	1.21

Appendix table 4.2
Pesticide use in soybeans, 1960-2008

	Millions of planted acres	Pounds active ingredient (a.i.) per planted acre and total pounds a.i. of pesticide applied, by type							
		Herbicide	Insecticide	Fungicide	Total	Herbicide	Insecticide	Fungicide	Total
		Millions of pounds a.i. applied				Pounds a.i. per planted acre			
1997	70.0	87.15	1.22	*	89.26	1.24	0.02	ŧ	1.28
1998	72.0	81.28	0.21	NA	81.49	1.13	0.00	NA	1.13
1999	73.7	78.62	0.29	NA	78.91	1.07	0.00	NA	1.07
2000	74.3	80.81	0.03	NA	80.84	1.09	0.00	NA	1.09
2001	74.1	78.78		NA	78.78	1.06	NA	NA	1.06
2002	74.0	86.23	0.54	*	86.79	1.17	0.01	ŧ	1.17
2003	73.4	88.00		NA	88.00	1.20	NA	NA	1.20
2004	75.2	88.41	0.60	*	89.03	1.18	0.01	ŧ	1.18
2005	72.0	88.98	1.48	*	90.96	1.24	0.02	ŧ	1.26
2006	75.5	106.06	2.58	*	109.12	1.40	0.03	ŧ	1.44
2007	64.7	82.25	2.56	*	85.48	1.27	0.04	ŧ	1.32
2008	75.7	106.22	4.70	1.04	111.96	1.40	0.06	ŧ	1.48

*Indicates that less than 1 million pounds were applied. ŧ indicates less than 0.01. NA indicates not available.
The total columns include other pesticide types, such as defoliants and desiccants.

Source: Economic Research Service with USDA and proprietary data. See Appendix 2.

1996-2008: Adoption of HT soybeans increases glyphosate use and reduces tillage

HT seeds were commercially introduced in 1996. Initially, only 7 percent of soybean acres were planted with HT seeds, but adoption rates increased rapidly. Herbicide use on soybeans initially declined. Only 1.07 pounds of herbicide were applied per acre in 1999, the lowest in over 25 years (app. fig. 4.2 and app. table 4.2). By 2008, however, usage had risen to 1.4 pounds per planted acre. By 2008, HT seeds were planted on over 90 percent of soybean acres, and glyphosate accounted for over 85 percent of total pounds applied (app. fig. 4.9). This heavy reliance on glyphosate for weed control prompted concern about resistance and encouraged soybean growers to look for strategies to mitigate that resistance (Owen, 2010).

With the advent of the soybean aphid in 2000 in the Midwest, some analysts argued that more soybean acres should be treated with insecticides (Gianessi, 2009; Ragsdale et al., 2011). Insecticide use on soybeans grew from just over 1 million pounds in 2005 to nearly 5 million pounds in 2008 (app. table 4.2), whereas insecticide use on corn and cotton declined over the same period. By 2008, the quantity of insecticides used on soybeans exceeded that on corn.

Reports of Asian soybean rust in the United States in 2004 and subsequent years encouraged preventative use of fungicides. Foliar fungicides were applied to 1 million acres of soybeans in 2008, amounting to 1 million pounds (app. table 4.2).

Cotton Pesticide Trends

Pesticide use on cotton increased from 77 million pounds active ingredient in 1960 to 143 million pounds in 1972. Usage declined to 34 million pounds in 1985, increased to 87 million pounds in 2000, and declined again to 38 million pounds in 2008 (app. table 4.3). In 2008, pesticide use in cotton production accounted for 7 percent of total pounds applied to the major crops, and about one-sixth of the quantity used in corn production.

1960-1981: Herbicide use increases rapidly, but changes in insecticide active ingredients reduce total insecticide use

Insecticides were widely used on cotton acres during the 1960s. However, due to environmental concerns, regulatory actions, and the introduction of new insecticides, usage began to decline in the 1970s. Insecticide use increased from 67 million pounds a.i. in 1960 to 105 million pounds in 1972, before declining to 19 million pounds in 1980. Cotton farmers replaced organochlorines, such as DDT and toxaphene, with other insecticides. Farmers also adopted new cropping practices (for instance, rotation with sorghum and other crops) to control insect infestations. Notably, DDT and toxaphene together accounted for about 50 percent of cotton pesticide use in 1968 (app. fig. 4.10). However, by the mid-1970s, DDT was no longer used, and the top insecticides applied to cotton were methyl-parathion and toxaphene (Eichers et al., 1978). Also, synthetic pyrethroids (e.g., permethrin, cypermethrin) were introduced and became widely used (Osteen and Fernandez-Cornejo, 2013). Synthetic pyrethroids are applied at much lower rates than older insecticides—fractions of an ounce compared with several pounds per acre.

As a result, the quantity of insecticides used declined from 62 million pounds in 1977 to 19 million pounds in 1980, or from 4.5 pounds per planted cotton acre to 1.3 pounds (app. fig. 4.3). Herbicides were used far less intensively than insecticides during this period. Nonetheless, driven by increases in trifluralin and fluometuron use (Andrilenas, 1974; Eichers et al., 1970 and 1978), herbicide use per planted acre rose from less than 0.1 pound in the early 1960s to 1.9 pounds in 1976 (app. table 4.3).

Appendix figure 4.3
Pounds of pesticide active ingredient (a.i.) per planted acre, cotton, 1960-2008

Sources: Economic Research Service with USDA and proprietary data. See Appendix 2.

Pesticide Use in U.S. Agriculture: 21 Selected Crops, 1960-2008, EIB-124
Economic Research Service/USDA

Appendix table 4.3
Pesticide use in cotton, 1960-2008

		Pounds active ingredient (a.i.) per planted acre and total pounds a.i. of pesticide applied, by type									
	Millions of planted acres	Herbicide	Insecticide	Fungicide	Other	Total	Herbicide	Insecticide	Fungicide	Other	Total
		Millions of pounds a.i. applied					*Pounds a.i. per planted acre*				
1960	16.1	1.39	67.11	*	8.04	76.54	0.09	4.17	ŧ	0.50	4.76
1961	16.6	1.17	76.12	*	9.00	86.29	0.07	4.59	ŧ	0.54	5.20
1962	16.3	1.48	77.47	*	9.96	88.91	0.09	4.75	ŧ	0.61	5.46
1963	14.8	1.65	78.73	*	10.92	91.37	0.11	5.30	ŧ	0.74	6.16
1964	14.8	2.01	67.90	*	11.88	81.96	0.14	4.58	ŧ	0.80	5.52
1965	14.2	5.39	58.77	*	12.84	77.27	0.38	4.15	ŧ	0.91	5.46
1966	10.3	5.48	62.94	*	13.80	82.60	0.53	6.08	ŧ	1.33	7.98
1967	9.4	5.96	89.55	*	14.92	110.78	0.63	9.48	ŧ	1.58	11.72
1968	10.9	13.61	76.73	*	16.04	106.71	1.25	7.03	ŧ	1.47	9.78
1969	11.9	12.80	52.73	*	17.17	82.97	1.08	4.44	ŧ	1.44	6.98
1970	11.9	15.23	63.28	*	18.29	97.05	1.28	5.30	ŧ	1.53	8.12
1971	12.4	20.97	78.64	*	19.41	119.24	1.70	6.37	ŧ	1.57	9.65
1972	14.0	19.53	105.10	*	17.99	142.80	1.40	7.51	ŧ	1.28	10.20
1973	12.5	24.28	67.40	*	16.57	108.39	1.95	5.40	ŧ	1.33	8.69
1974	13.7	21.13	72.15	*	15.14	108.54	1.54	5.27	ŧ	1.11	7.93
1975	9.5	16.01	46.80	*	13.72	76.62	1.69	4.94	ŧ	1.45	8.08
1976	11.6	21.91	69.17	*	12.30	103.43	1.88	5.94	ŧ	1.06	8.89
1977	13.7	19.77	62.10	*	11.38	93.33	1.45	4.54	ŧ	0.83	6.82
1978	13.4	20.01	21.16	*	10.47	51.74	1.50	1.58	ŧ	0.78	3.87
1979	14.0	20.31	19.01	*	9.55	49.00	1.45	1.36	ŧ	0.68	3.51
1980	14.5	21.69	18.72	*	8.63	49.19	1.49	1.29	ŧ	0.59	3.38
1981	14.3	24.17	24.41	*	7.72	56.47	1.69	1.70	ŧ	0.54	3.94
1982	11.3	19.29	19.10	*	6.80	45.39	1.70	1.68	ŧ	0.60	4.00
1983	7.9	15.75	15.92	*	6.26	38.12	1.99	2.01	ŧ	0.79	4.81
1984	11.1	20.10	12.10	*	5.72	38.09	1.80	1.09	ŧ	0.51	3.42
1985	10.7	16.32	12.45	*	5.18	34.10	1.53	1.17	ŧ	0.49	3.19
1986	10.0	20.83	15.74	*	4.64	41.34	2.07	1.57	ŧ	0.46	4.12
1987	10.4	19.46	15.27	*	4.11	38.94	1.87	1.47	ŧ	0.39	3.75
1988	12.5	21.64	17.48	*	3.57	42.77	1.73	1.40	ŧ	0.29	3.42
1989	10.6	18.73	12.37	*	3.03	34.20	1.77	1.17	ŧ	0.29	3.23
1990	12.3	25.27	10.10	*	2.49	37.90	2.05	0.82	ŧ	0.20	3.07
1991	14.1	29.20	9.73	*	11.44	50.82	2.08	0.69	ŧ	0.81	3.62
1992	13.2	27.99	15.20	*	12.04	55.78	2.11	1.15	ŧ	0.91	4.21
1993	13.4	29.24	23.48	*	12.58	66.44	2.18	1.75	ŧ	0.94	4.94
1994	13.7	33.88	23.94	*	11.60	70.29	2.47	1.75	ŧ	0.85	5.12
1995	16.9	40.05	28.21	*	16.80	85.88	2.37	1.67	ŧ	0.99	5.07
1996	14.7	31.06	23.56	*	26.11	81.71	2.12	1.61	ŧ	1.78	5.58

Pesticide Use in U.S. Agriculture: 21 Selected Crops, 1960-2008, EIB-124
Economic Research Service/USDA

Pesticide use in cotton, 1960-2008—continued

	Millions of planted acres	Herbicide	Insecticide	Fungicide	Other	Total	Herbicide	Insecticide	Fungicide	Other	Total
				Millions of pounds a.i. applied					Pounds a.i. per planted acre		
1997	13.9	30.84	23.30	1.29	21.53	76.97	2.22	1.68	0.09	1.55	5.54
1998	13.4	26.43	15.94	*	10.96	53.81	1.97	1.19	ŧ	0.82	4.02
1999	14.9	30.48	40.13	*	13.14	84.55	2.05	2.70	ŧ	0.88	5.68
2000	15.5	32.62	41.25	*	12.70	87.16	2.10	2.66	ŧ	0.82	5.62
2001	15.8	33.09	24.92	*	14.03	72.62	2.10	1.58	ŧ	0.89	4.61
2002	14.0	34.35	6.31	*	14.28	55.35	2.46	0.45	ŧ	1.02	3.97
2003	13.5	29.95	13.92	*	13.33	57.54	2.22	1.03	ŧ	0.99	4.27
2004	13.7	34.11	9.10	*	15.79	59.25	2.50	0.67	ŧ	1.16	4.34
2005	14.2	31.91	15.23	*	15.71	63.06	2.24	1.07	ŧ	1.10	4.43
2006	15.3	34.67	8.41	*	17.88	61.16	2.27	0.55	ŧ	1.17	4.00
2007	10.8	26.09	6.62	*	13.67	46.44	2.41	0.61	ŧ	1.26	4.29
2008	9.5	22.43	5.32	*	9.75	37.56	2.37	0.56	ŧ	1.03	3.97

*Indicates that less than 1 million pounds were applied. ŧ indicates less than 0.01.

Source: Economic Research Service with USDA and proprietary data. See Appendix 2.

1982-1995: Fluctuation in pesticide use and changes in active ingredients used

In the 1970s, cotton farmers controlled late-season pests with synthetic pyrethroids (e.g., permethrin). However, by the 1980s, tobacco budworm resistance to pyrethroids induced growers to increase the use of organophosphates such as methyl parathion and azinphos-methyl (Morgan, 2002; Texas AgriLife Extension Services, 2009), adopt new IPM strategies, and plant new early maturing cotton varieties (app. fig. 4.2).

Starting in the 1970s, cotton growers and State governments cooperated to eradicate the boll weevil. Different cotton growing regions joined the program in different years. Typically, the first year of a program entailed heavy application of pesticides, generally malathion. In subsequent years, the boll weevil population was monitored and treated as needed. A new wave of cotton producing regions began participation in 1993. By 1995, insecticide use increased to 28 million pounds, the highest since 1977 (app. table 4.3). Herbicide use ranged from 1.5 to 2.5 pounds per planted acre in most years from 1982 to 1995.

1996-2008: Adoption of GE cotton reduces insecticide use and increases glyphosate use

The adoption of Bt cotton since 1996 and the reduction of insecticide use for boll weevil eradication after 2000 contributed to reductions in overall insecticide use. Bt cotton grew from 15 percent of total acres in 1996 to more than 60 percent by 2008. Insecticide use was more than 1 pound per planted acre in most years before the introduction of Bt cotton but declined to 0.6 pound by 2008 (app. table 4.3).

Driven primarily by malathion use for boll weevil eradication programs, insecticide use increased to over 2 pounds per planted acre in both 1999 and 2000 (USDA/NASS, *Agricultural Chemical Usage: Field Crop Summary,* 2000 and 2001). Total insecticide use on cotton exceeded 40 million pounds in these years, the highest since 1977, before declining to 5 million pounds by 2008. Acephate accounted for the largest quantity of insecticide used in 2008, almost 6 percent of cotton pesticide use (app. fig. 4.11).

Herbicide-tolerant cotton was commercially introduced in 1996. HT seeds were quickly accepted by cotton farms and accounted for approximately 10 percent of cotton planted acres after 1 year. As with other crops, HT adoption had a profound impact on herbicide use. In 1996, trifluralin, MSMA (monosodium methanearsonate), pendimethalin, and fluometuron were dominant herbicides in cotton production (USDA/NASS, *Agricultural Chemical Usage: 1996 Field Crop Summary,* 1997). By 2008, as HT cotton seeds were planted on 70 percent of cotton acres, glyphosate use accounted for 37 percent of cotton pesticide use (app. fig. 4.11). As HT cotton was adopted, total herbicide use per cotton acre fell (app. table 3.3), though it increased to 2.4 pounds per acre in 2007 and 2008, possibly due to the development of weed resistance to glyphosate (Owen, 2010). "Other pesticides" (desiccants, defoliants, and growth regulators) became the second largest pesticide type applied to cotton after 2002 (app. table 3.3). Tribufos and ethephon were among the top 10 cotton pesticides in 2008 (appendix figure 4.11).

Potato Pesticide Trends

Pesticide use on fall potatoes increased from 9 million pounds a.i. in 1960 to 60 million pounds in 1999 before declining to 53 million pounds in 2008 (app. table 4.4). Other pesticides, including fumigants and desiccants, constituted 44 million of the 53 million pounds applied to potatoes in 2008. Of the 9 million pounds remaining, 2.3 million pounds were herbicides, 1.4 million pounds were insecticides, and 5 million pounds were fungicides. Fall potatoes—a leading vegetable crop

Appendix figure 4.4
Pounds of pesticide active ingredient (a.i.) per planted acre, potatoes, 1960-2008

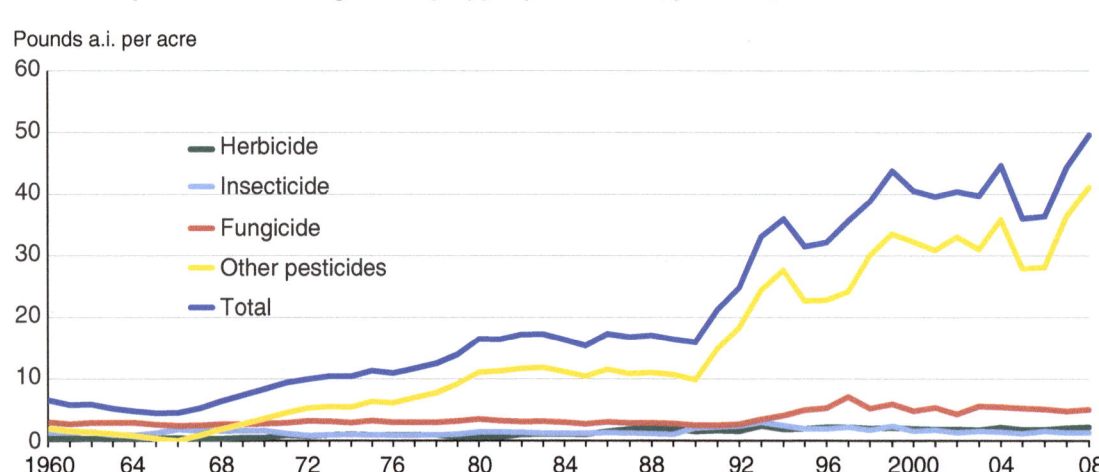

Sources: Economic Research Service with USDA and proprietary data. See Appendix 2.

Appendix table 4.4
Pesticide use in potatoes, 1960-2008

		Pounds active ingredient (a.i.) per planted acre and total pounds a.i. of pesticide applied, by type									
	Millions of planted acres	Herbicide	Insecticide	Fungicide	Other	Total	Herbicide	Insecticide	Fungicide	Other	Total
		Millions of pounds a.i. applied					Pounds a.i. per planted acre				
1960	14.0	*	1.89	4.10	2.71	9.20	0.35	1.35	2.93	1.94	6.57
1961	15.0	*	1.95	4.01	2.26	8.75	0.35	1.29	2.65	1.50	5.79
1962	14.0	*	1.83	3.91	1.81	8.00	0.33	1.33	2.84	1.31	5.81
1963	13.0	*	1.28	3.82	1.36	6.91	0.34	0.96	2.85	1.02	5.17
1964	13.0	*	1.09	3.72	0.91	6.19	0.35	0.83	2.84	0.69	4.72
1965	14.0	*	1.67	3.63	0.46	6.30	0.39	1.18	2.56	0.32	4.44
1966	15.0	*	2.63	3.53	0.01	6.73	0.37	1.76	2.36	0.01	4.49
1967	15.0	*	2.40	3.65	1.29	7.83	0.33	1.60	2.43	0.86	5.23
1968	14.0	*	2.22	3.77	2.57	8.98	0.31	1.57	2.66	1.81	6.35
1969	15.0	*	2.29	3.88	3.84	10.67	0.45	1.57	2.66	2.64	7.32
1970	14.0	*	2.32	4.00	5.12	12.10	0.45	1.60	2.76	3.53	8.34
1971	14.0	1.25	1.64	4.12	6.40	13.41	0.87	1.14	2.88	4.47	9.36
1972	13.0	*	1.05	4.13	6.84	12.95	0.72	0.81	3.17	5.25	9.95
1973	13.0	1.24	1.22	4.14	7.27	13.87	0.94	0.91	3.11	5.47	10.43
1974	14.0	1.52	1.41	4.15	7.71	14.79	1.07	0.99	2.92	5.42	10.40
1975	13.0	1.22	1.13	4.16	8.14	14.65	0.94	0.87	3.20	6.27	11.28
1976	14.0	1.35	1.26	4.17	8.58	15.37	0.96	0.90	2.97	6.11	10.94
1977	14.0	1.28	1.20	4.15	9.68	16.30	0.91	0.86	2.96	6.92	11.65
1978	14.0	1.30	1.35	4.12	10.78	17.55	0.93	0.96	2.94	7.70	12.53
1979	13.0	*	1.39	4.10	11.89	18.10	0.56	1.07	3.17	9.17	13.97
1980	12.0	*	1.63	4.08	12.99	19.35	0.56	1.39	3.47	11.05	16.47
1981	13.0	*	1.73	4.05	14.09	20.59	0.57	1.38	3.23	11.22	16.40
1982	13.0	1.36	1.77	4.03	15.19	22.34	1.04	1.36	3.09	11.66	17.15
1983	13.0	1.40	1.54	3.97	15.01	21.92	1.10	1.21	3.12	11.81	17.24
1984	13.0	1.44	1.59	3.91	14.83	21.78	1.08	1.20	2.93	11.12	16.33
1985	14.0	1.49	1.68	3.84	14.65	21.66	1.06	1.19	2.73	10.42	15.40
1986	13.0	1.90	1.60	3.78	14.47	21.76	1.51	1.27	3.01	11.52	17.31
1987	13.0	2.45	1.63	3.72	14.30	22.10	1.86	1.24	2.83	10.86	16.78
1988	13.0	2.66	1.43	3.66	14.12	21.87	2.07	1.12	2.85	10.99	17.02
1989	13.0	2.45	1.40	3.60	13.94	21.39	1.88	1.07	2.76	10.68	16.39
1990	14.0	2.14	2.89	3.54	13.76	22.32	1.53	2.06	2.53	9.83	15.95
1991	14.0	2.35	3.12	3.47	20.93	29.87	1.67	2.22	2.46	14.87	21.22
1992	13.0	2.06	3.21	3.57	24.44	33.28	1.54	2.40	2.66	18.25	24.85
1993	14.0	3.26	4.07	4.77	33.84	46.00	2.39	2.93	3.43	24.35	33.10
1994	14.0	2.64	3.48	5.77	39.23	51.13	1.86	2.45	4.06	27.59	35.96
1995	14.0	2.70	2.69	6.92	31.76	44.07	1.93	1.92	4.94	22.68	31.46
1996	15.0	3.18	2.84	7.65	33.12	46.79	2.18	1.96	5.26	22.77	32.17

Pesticide Use in U.S. Agriculture: 21 Selected Crops, 1960-2008, EIB-124
Economic Research Service/USDA

Pesticide use in potatoes, 1960-2008—continued

	Millions of planted acres	Pounds active ingredient (a.i.) per planted acre and total pounds a.i. of pesticide applied, by type									
		Herbicide	Insecticide	Fungicide	Other	Total	Herbicide	Insecticide	Fungicide	Other	Total
		Millions of pounds a.i. applied					*Pounds a.i. per planted acre*				
1997	14.0	3.05	3.03	9.79	33.47	49.33	2.21	2.19	7.08	24.19	35.66
1998	14.0	2.81	2.44	7.39	42.37	55.01	1.98	1.72	5.22	29.93	38.85
1999	14.0	2.80	3.23	8.08	46.13	60.24	2.03	2.35	5.87	33.52	43.77
2000	14.0	2.69	2.14	6.62	44.60	56.06	1.95	1.55	4.79	32.25	40.53
2001	12.0	2.25	2.12	6.62	38.42	49.40	1.80	1.70	5.31	30.81	39.62
2002	13.0	2.40	1.65	5.54	42.85	52.44	1.85	1.27	4.26	32.97	40.35
2003	13.0	2.14	1.95	6.99	39.45	50.53	1.68	1.53	5.49	30.97	39.67
2004	12.0	2.54	1.65	6.42	42.66	53.26	2.13	1.38	5.38	35.77	44.67
2005	11.0	1.97	1.26	5.76	30.90	39.89	1.77	1.14	5.19	27.88	35.99
2006	11.0	2.03	1.74	5.77	31.91	41.44	1.78	1.53	5.06	28.00	36.37
2007	11.0	2.28	1.45	5.43	41.49	50.66	2.00	1.27	4.76	36.33	44.36
2008	11.0	2.32	1.38	5.31	43.52	52.53	2.19	1.30	5.01	41.07	49.58

*Indicates that less than 1 million pounds were applied. t indicates less than 0.01.
Source: Economic Research Service with USDA and proprietary data. See Appendix 2.

in the United States, contributing to 18 percent of vegetable sales in 2008 (USDA/ERS, 2010a)—accounted for 10 percent of pesticide use in 2008.

While herbicide use in potatoes as measured in pounds is small, most potato acres are treated with herbicides. Metribuzin, pendimethalin, and rimsulfuron are the most commonly used herbicides in fall potato production. In 2005, metribuzin was applied to 74 percent of acres; pendimethalin and rimsulfuron were each applied to about 30 percent of this acreage (USDA/NASS, *Agricultural Chemical Usage,* various years). Commonly used pre-emergence herbicides include EPTC (since the 1960s) and metolachlor (since the 1990s) (Lin et al., 1995). EPTC, applied at high rates, has dominated in pounds applied.

The insecticides DDT and malathion were among the top pesticides used on potatoes in the 1960s and disulfoton after (Lin et al., 1995). In recent years, the top pesticides applied to potatoes were the fumigants metam sodium, dichloropropene, and metam potassium followed by the multi-site fungicides mancozeb and chlorothalonil (USDA/NASS, *Agricultural Chemical Usage,* various years).

Wheat Pesticide Trends

Pesticide use on wheat increased from 6 million pounds a.i. in 1960 to a peak of 31 million pounds in 1980 (app. table 4.5). Since then, use has ranged from 14 million pounds in 1992 to 23 million pounds in 2008 (app. table 4.5). Even though wheat occupies the third largest acreage of U.S. field crops (after corn and soybeans), pesticide use in wheat production accounted for less than 5 percent of pesticides applied to the 21 crops in 2008. Generally, per-acre pesticide use is greater on spring and durum wheat than on winter wheat, due in large part to herbicide use; herbicides are applied to 30-60 percent of winter wheat acres, versus 88-100 percent of durum wheat acres (USDA/NASS,

Pounds of pesticide active ingredient (a.i.) per planted acre, wheat, 1960-2008

Pounds a.i. per acre

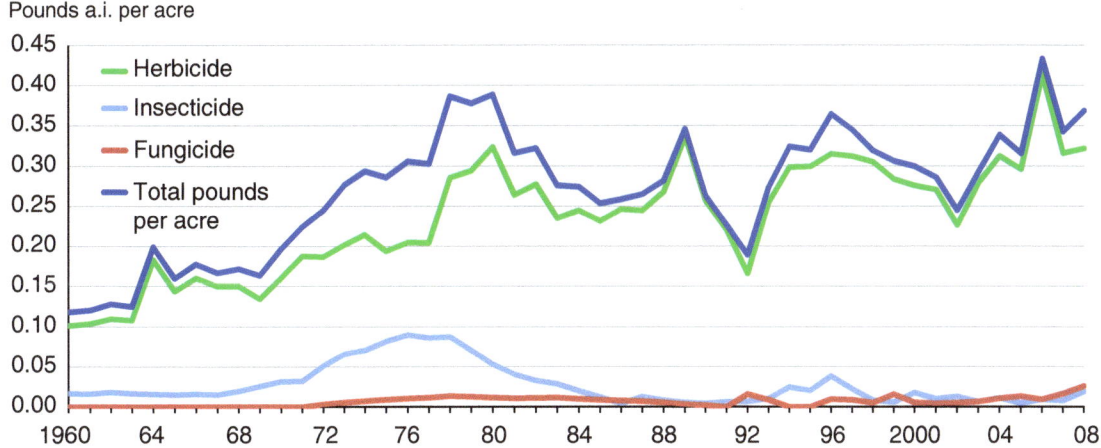

Sources: Economic Research Service with USDA and proprietary data. See Appendix 2.

Agricultural Chemical Usage, various years). Usually, less than 10 percent of wheat acres are treated with insecticides or foliar fungicides, so less than 0.1 pound of insecticides or fungicides were applied per planted acre from 1960 to 2008.

1960-1981: Rapid increase in pesticide use

During the 1960s and 1970s, herbicides replaced tillage and cultivation practices as primary weed controls on wheat. Herbicide use increased from 0.1 pound per planted acre in 1960 to 0.32 pound acre in 1980 (app. figure 4.5, app. table 4.5). 2,4-D accounted for most wheat pesticide use from 1960 through 1971 (Lin et al., 1995). By 1971, nearly 9 million pounds of 2,4-D were applied (Andrilenas, 1974). Other commonly used broadleaf herbicides included MCPA and dicamba. Triallate was introduced in 1972.

1982-2008: Fluctuation in pesticide use and changes in active ingredients used

Herbicide use ranged between 0.2 and 0.3 pound per planted acre in most years (app. fig. 4.5 and app. table 4.5). 2,4-D, MCPA, and dicamba were used extensively during this period (USDA/NASS, *Agricultural Chemical Usage,* various years). No-till practices in the winter wheat areas boosted glyphosate use (Thompson, 2011). In recent years, glyphosate—followed by 2,4-D, MPCA, dicamba, and bromoxynil—was the top pesticide used on wheat (USDA/NASS, *Agricultural Chemical Usage,* various years).

Pesticide Use in U.S. Agriculture: 21 Selected Crops, 1960-2008, EIB-124
Economic Research Service/USDA

Pesticide use in wheat, 1960-2008

	Millions of planted acres	Pounds active ingredient (a.i.) per planted acre and total pounds a.i. of pesticide applied, by type							
		Herbicide	Insecticide	Fungicide	Total	Herbicide	Insecticide	Fungicide	Total
		Millions of pounds a.i. applied				Pounds a.i. per acre			
1960	54.9	5.56	*	*	6.48	0.10	ŧ	ŧ	0.12
1961	55.7	5.81	*	*	6.72	0.10	ŧ	ŧ	0.12
1962	49.3	5.41	*	*	6.31	0.11	0.02	ŧ	0.13
1963	53.4	5.78	*	*	6.68	0.11	0.02	ŧ	0.13
1964	55.7	10.20	*	*	11.09	0.18	0.02	ŧ	0.20
1965	57.4	8.28	*	*	9.19	0.14	0.02	ŧ	0.16
1966	54.1	8.68	*	*	9.61	0.16	0.02	ŧ	0.18
1967	67.3	10.10	1.04	*	11.23	0.15	0.02	0.02	0.17
1968	61.9	9.29	1.21	*	10.63	0.15	0.02	0.02	0.17
1969	53.5	7.19	1.38	*	8.74	0.13	0.03	0.03	0.16
1970	48.7	7.85	1.54	*	9.60	0.16	0.03	0.03	0.20
1971	53.8	10.11	1.71	*	12.07	0.19	0.03	0.03	0.22
1972	54.9	10.25	2.82	*	13.44	0.19	0.05	0.05	0.24
1973	59.3	11.95	3.92	*	16.36	0.20	0.07	0.07	0.28
1974	71.0	15.22	5.03	*	20.86	0.21	0.07	0.07	0.29
1975	74.9	14.55	6.13	*	21.42	0.19	0.08	0.08	0.29
1976	80.4	16.48	7.24	*	24.57	0.20	0.09	0.09	0.31
1977	75.4	15.41	6.51	*	22.80	0.20	0.09	0.09	0.30
1978	66.0	18.83	5.78	*	25.51	0.29	0.09	0.09	0.39
1979	71.4	21.01	5.04	*	26.98	0.29	0.07	0.07	0.38
1980	80.8	26.15	4.31	*	31.41	0.32	0.05	0.05	0.39
1981	88.3	23.34	3.58	*	27.90	0.26	0.04	0.04	0.32
1982	86.2	23.92	2.85	1.00	27.78	0.28	0.03	0.03	0.32
1983	76.4	17.97	2.22	*	21.09	0.24	0.03	0.03	0.28
1984	79.2	19.36	1.58	*	21.74	0.24	0.02	0.02	0.27
1985	75.5	17.50	*	*	19.13	0.23	0.01	ŧ	0.25
1986	72.0	17.75	*	*	18.64	0.25	0.00	ŧ	0.26
1987	65.8	16.13	*	*	17.45	0.24	0.01	ŧ	0.27
1988	65.5	17.54	*	*	18.47	0.27	0.01	ŧ	0.28
1989	76.6	25.80	*	*	26.51	0.34	0.01	ŧ	0.35
1990	77.0	19.81	*	*	20.31	0.26	0.00	ŧ	0.26
1991	69.9	15.39	*	*	15.86	0.22	0.01	ŧ	0.23
1992	72.2	12.01	*	1.16	13.66	0.17	0.01	ŧ	0.19
1993	72.2	18.38	*	*	19.77	0.25	0.01	ŧ	0.27
1994	70.3	21.01	1.77	*	22.80	0.30	0.03	0.03	0.32
1995	69.0	20.66	1.43	*	22.11	0.30	0.02	0.02	0.32
1996	75.1	23.67	2.89	*	27.37	0.32	0.04	0.04	0.36
1997	70.4	21.99	1.58	*	24.28	0.31	0.02	0.02	0.34

Pesticide Use in U.S. Agriculture: 21 Selected Crops, 1960-2008, EIB-124
Economic Research Service/USDA

Appendix table 4.5

Pesticide use in wheat, 1960-2008—continued

| | | **Pounds active ingredient (a.i.) per planted acre and total pounds a.i. of pesticide applied, by type** | | | | | | | |
	Millions of planted acres	Herbicide	Insecticide	Fungicide	Total	Herbicide	Insecticide	Fungicide	Total
		Millions of pounds a.i. applied				*Pounds a.i. per acre*			
1998	65.8	20.09	*	*	20.99	0.31	0.01	ŧ	0.32
1999	62.7	17.80	*	*	19.19	0.28	0.01	ŧ	0.31
2000	62.5	17.24	1.14	*	18.74	0.28	0.02	0.02	0.30
2001	59.4	16.10	*	*	17.03	0.27	0.01	ŧ	0.29
2002	60.3	13.66	*	*	14.78	0.23	0.01	ŧ	0.25
2003	62.1	17.35	*	*	18.21	0.28	0.01	ŧ	0.29
2004	59.6	18.65	*	*	20.22	0.31	0.01	ŧ	0.34
2005	57.2	16.95	*	*	18.06	0.30	0.01	ŧ	0.32
2006	57.3	23.76	*	*	24.88	0.41	0.01	ŧ	0.43
2007	60.5	19.12	*	1.02	20.74	0.32	0.01	ŧ	0.34
2008	63.2	20.37	1.26	1.68	23.31	0.32	0.02	0.02	0.37

*Indicates that less than 1 million pounds were applied. ŧ indicates less than 0.01.
The total columns include other pesticide types, such as defoliants and desiccants.

Source: Economic Research Service with USDA and proprietary data. See Appendix 2.

Pesticide Use in U.S. Agriculture: 21 Selected Crops, 1960-2008, EIB-124
Economic Research Service/USDA

Appendix figure 4.6
Pesticide use by active ingredient (a.i.) on corn in 1968, percent of total pounds a.i. applied[1]

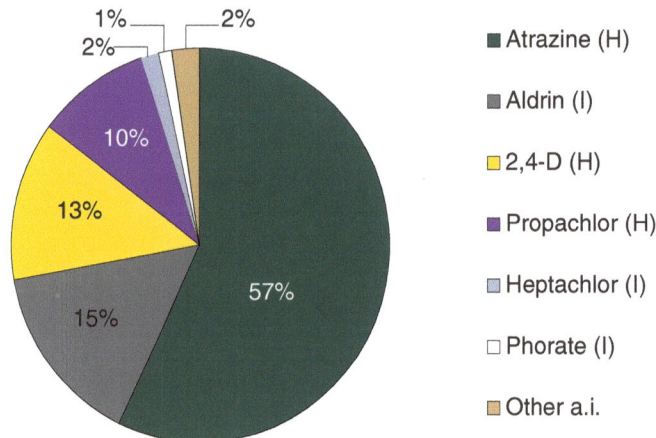

1% 2%
2%
10%
13%
15%
57%

- Atrazine (H)
- Aldrin (I)
- 2,4-D (H)
- Propachlor (H)
- Heptachlor (I)
- Phorate (I)
- Other a.i.

[1]This graph shows the top pesticide a.i. used in corn in 1968.
Sources: Economic Research Service with USDA and proprietary data. See Appendix 2.

Appendix figure 4.7
Pesticide use by active ingredient (a.i.) on corn in 2008, percent of total pounds a.i. applied[1]

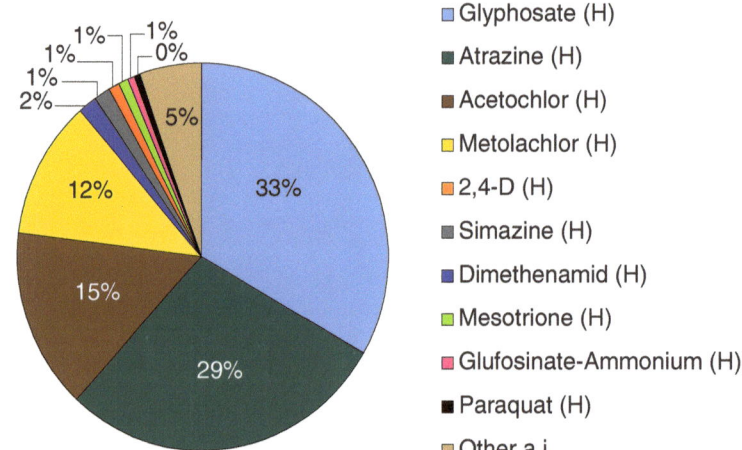

1% 1%
1% 0%
1%
1%
2%
5%
12%
15%
33%
29%

- Glyphosate (H)
- Atrazine (H)
- Acetochlor (H)
- Metolachlor (H)
- 2,4-D (H)
- Simazine (H)
- Dimethenamid (H)
- Mesotrione (H)
- Glufosinate-Ammonium (H)
- Paraquat (H)
- Other a.i.

[1]This graph shows the top pesticide a.i. used in corn in 2008.
Sources: Economic Research Service with USDA and proprietary data. See Appendix 2.

Pesticide Use in U.S. Agriculture: 21 Selected Crops, 1960-2008, EIB-124
Economic Research Service/USDA

Appendix figure 4.8
Pesticide use by active ingredient (a.i.) on soybeans in 1968, percent of total pounds a.i. applied[1]

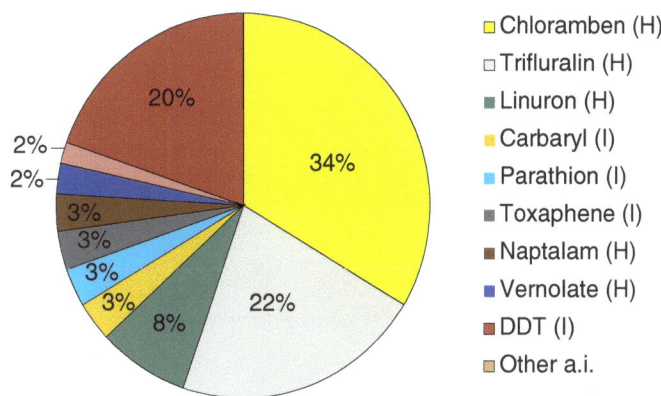

- Chloramben (H)
- Trifluralin (H)
- Linuron (H)
- Carbaryl (I)
- Parathion (I)
- Toxaphene (I)
- Naptalam (H)
- Vernolate (H)
- DDT (I)
- Other a.i.

[1]This graph shows the top pesticide a.i. used in soybeans in 1968.
Sources: Economic Research Service with USDA and proprietary data. See Appendix 2.

Appendix figure 4.9
Pesticide use by active ingredient (a.i.) on soybeans in 2008, percent of total pounds a.i. applied[1]

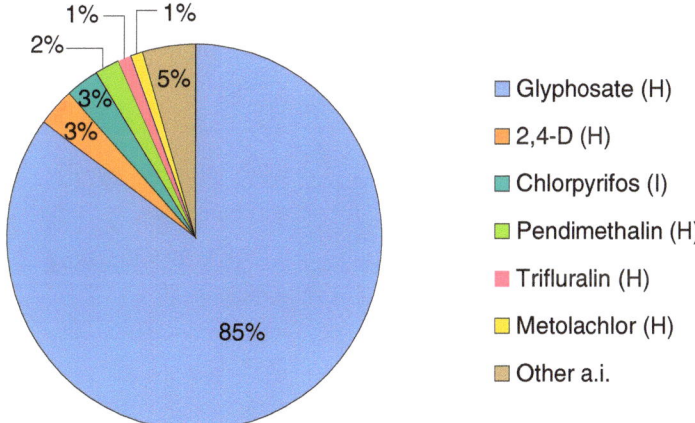

- Glyphosate (H)
- 2,4-D (H)
- Chlorpyrifos (I)
- Pendimethalin (H)
- Trifluralin (H)
- Metolachlor (H)
- Other a.i.

[1]This graph shows the top pesticide a.i. used in soybeans in 2008.
Sources: Economic Research Service with USDA and proprietary data. See Appendix 2.

Pesticide Use in U.S. Agriculture: 21 Selected Crops, 1960-2008, EIB-124
Economic Research Service/USDA

Appendix figure 4.10
Pesticide use by active ingredient (a.i.) on cotton in 1968, percent of total pounds a.i. applied[1]

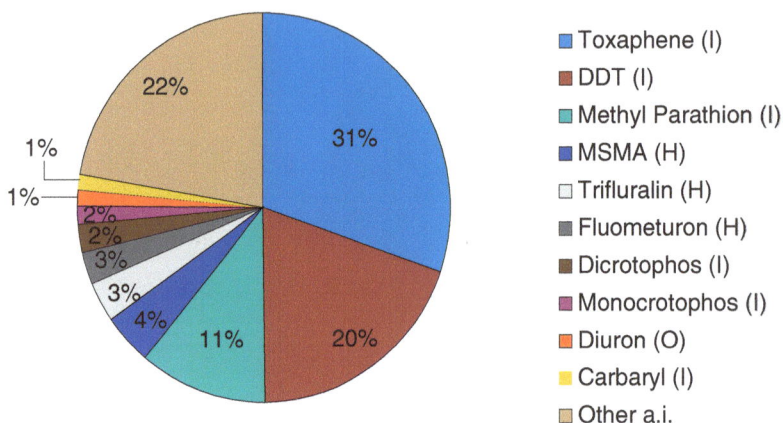

[1]This graph shows the top pesticide a.i. used in soybeans in 1968.
Sources: Economic Research Service with USDA and proprietary data. See Appendix 2.

Appendix figure 4.11
Pesticide use by active ingredient (a.i.) on cotton in 2008, percent of total pounds a.i. applied[1]

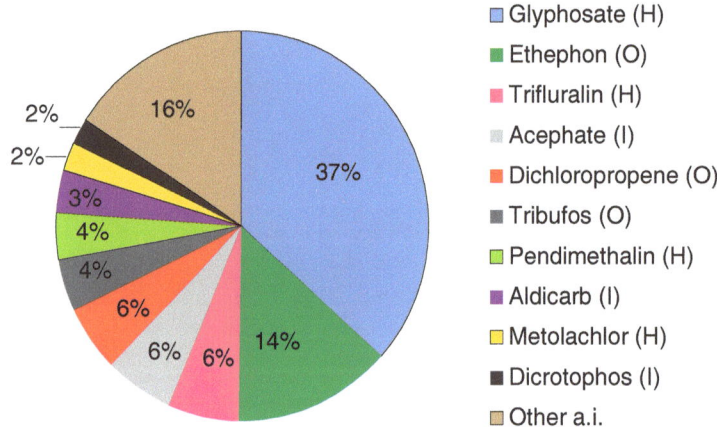

[1]This graph shows the top pesticide a.i. used in cotton in 2008.
Sources: Economic Research Service with USDA and proprietary data. See Appendix 2.